The Universal Subject of Our Time

(or: How I Learned to Stop Worrying and Love the Machine)

Darius Nikbin

Winchester, UK
Washington, USA

First published by Zero Books, 2019
Zero Books is an imprint of John Hunt Publishing Ltd., No. 3 East St., Alresford,
Hampshire SO24 9EE, UK
office1@jhpbooks.net
www.johnhuntpublishing.com
www.zero-books.net

For distributor details and how to order please visit the 'Ordering' section on our website.

ISBN: 978 1 78904 040 1
978 1 78904 041 8 (ebook)
Library of Congress Control Number: 2018932661

A CIP catalogue record for this book is available from the British Library.

Design: Stuart Davies

Printed and bound by CPI Group (UK) Ltd, Croydon, CR0 4YY, UK

We operate a distinctive and ethical publishing philosophy in
all areas of our business, from our global network of authors to
production and worldwide distribution.

The Universal Subject of Our Time

(or: How I Learned to Stop Worrying
and Love the Machine)

Contents

This book is dedicated to Alexandra

The Girl with the Lion Tattoo

in Reykjavik, Iceland.

Introduction

Man vs Machine

Whether we are based on carbon or on silicon makes no fundamental difference; we should each be treated with appropriate respect.
Arthur C. Clarke

When in the winter of 2006, NASA robots from Earth were scouring the surface of Mars for signs of life, the reigning world chess champion, Vladimir Kramnik, faced a computer opponent, Deep Fritz, in *The World Chess Challenge: Man vs Machine*. The venue was Bonn, Germany. The human grandmaster, Kramnik, had already drawn the first game against his machine adversary. Now, in the second game of six, the most likely outcome was another tied result. Despite this, Kramnik pushed aggressively for a win. Deep Fritz responded in turn. The game edged towards an uncertain result. Then, when the contest seemed once again destined for a draw, unexpectedly, with his thirty-fourth move, Kramnik made a calamitous error.

Susan Polgar, the chess commentator, later described it as "The blunder of the century." Kramnik, shocked, held his head in his hands. Mathias Feist, the operator of Deep Fritz, made the winning move in a complete state of disbelief. Kramnik was checkmated and the machine had taken the lead. During the ensuing press conference, Kramnik attempted to explain what had happened: "It was very strange, some kind of blackout. I was feeling well, I was playing well, I think I was pretty much *better*...I was not feeling tired, I think I was calculating well during the whole game... It's just very strange, I cannot explain it." Deep Fritz went on to defeat his weary human opponent by four games to two.

Responding to the result, Peter Newborn, Professor of

Computer Science at Montreal University, said that "The science is done!" Historically, this match signified the final moment when machines had prevailed over man in the ancient game of chess. To this day, following the 2006 encounter, human vs computer chess games have become an irrelevancy; and now, in an era when computer processors are constantly accelerating in their calculating ability, even the simplest smartphones can overwhelm the most respected chess grandmasters. The machine had arrived on the scene by incomprehensibly beating man in a game of his own invention. Deep Fritz achieved this goal through a combination of logic and processing ability. Man now stood in the shadow of his silicon progeny.

Today, we live in a world where silicon, the semi-conductor material computer chips are based on, has infiltrated almost every aspect of our lives. This element provides the mechanical support for the integrated electronic circuitry of microchips, the building blocks of computers and other electronic devices. Silicon-based technology has created new freedoms within the space of the internet. Products of human thought, such as the World Wide Web (WWW), developed in the 1980s by Tim Berners-Lee and Robert Cailleau, empowered by high-speed broadband, smartphones and the mobile web have replaced traditional networks of human communication. New connections have been created that span the entire planet. In the United Kingdom alone, there are 45 million internet users, accounting for 70 per cent of the whole population. An astonishing 99 per cent of young people in Britain (16-24 years old) browse the web. Another recent survey showed that two-thirds of the UK populace has smartphones with each person spending 2 hours online every day, on average. The trends are reflected globally with the population of smartphone-users reaching 2 billion in 2016. That is more than a quarter of the world's population connected to the digital silicon interface.

How has this come about? To answer this question, a new,

deeper scientific and philosophical understanding of the relation between man and machine is needed. The word 'machine' is defined by the *Oxford English Dictionary* as 'an apparatus using mechanical power [that] performs a particular task'. Technology, then, is defined as the application of mechanical science for human purposes such as communication or 'machinery and devices developed from scientific knowledge'. Another useful definition, that for a 'robot', describes a mechanical object made to carry out tasks that would otherwise need human effort. These statements may guide us; however, this book aims to move far beyond their comfort zone, to address questions concerning the mind, subjectivity, the self and technology. Of paramount concern is the question: can a machine think?

Thinking Machines

Some would answer negatively, arguing that machines are nothing more than the sum of what they are made of. Wires, chips and circuitry, complex connections, amounting to no more than artificially programmable silicon, an electronic arrangement of transistors, electrically stimulated and equally subject to the control of humans, who could always pull the plug on it all. They would argue that thinking, the activity of the mind, is the exclusive preserve of man, even if some basic intelligence can be found in the animal kingdom. According to this line of thinking, the idea that machines can have a mind is a human fantasy, the raw material for science-fiction authors and cinematic escapism: if a machine can think, then pigs can fly. Human mastery of the machine is total, they say, and their usefulness guarantees our survival.

The science of robotics and rapid expansion of personal technology has weakened the fortifications of this traditional position. The tangibility of technology in our everyday lives, the increasing sensory engagement of sight, sound and now touch, suggests that the mechanical has infiltrated our very thought processes. Our human organic connections to one another have been rerouted via silicon. The statistics above prove that we have become drastically more engaged with technology in our lives to accomplish tasks that would have previously required human attention. Computers enable us to sift through the internet for data, the fundamental unit of computer information. This can apparently help fill gaps in our knowledge, but when we ask a search engine a question, are we aware that we are rerouting our attention away from direct human-to-human knowledge, and investing a modicum of our thought into an electronic network of data and information?

Here, technology requires an investment of human thought, a product of our organic energetic activity, into an inert inorganic silicon-based network. For example, if you put a question to a web search engine, such as 'What is the meaning of life?', the engine responds with a new configuration of answers that only indirectly reflect the knowledge others have put on the web. This leads to a separation or gap in direct human relations between friends, family and co-workers. Furthermore, we are now equipped with camera smartphones that capture instants, so that we no longer even have to remember anything or cherish moments in our lives. Human memory is becoming defunct. In a short space of time, technological machinery has proven that every human thought process, reducible to symbols and images, could be transplanted to the electronic world. That technology could leave us like empty shells, reverberating nothing, other than bland emotional responses to external stimuli from an expanding electronic web of silicon, is a central motivation of this book.

In view of how fast this technological integration has occurred, we must also ask whether there is anything fundamentally different between man and machine? Many of the emotions that we experience are the same as those found in the animal kingdom, so emotions are not uniquely human. In *The Expression of the Emotions in Man and Animals* (1872) published 13 years after *The Origin of the Species,* Charles Darwin went so far as to suggest that humans and even lower animals express similar emotions. What separates us from animals is an absence of total reliance on instinct. Human self-awareness is far accentuated compared to animals. The question here, therefore, is whether machines can be self-aware? This book is also concerned with the issue of what separates humans from robots or intelligent machines. The starting point and focus for this investigation will be subjectivity or the sense of self. Then we may answer the big question of who we are (or at least, who do we *think* we are?).

Subjectivity: the Sense of Selfhood

What is subjectivity? Subjectivity comes into being through the senses in, for example, the act of seeing. In this book, the terms 'subject' and 'self' will be used interchangeably. Our sense of subjectivity, or self, forms the very basis of our being and enables us to create a boundary between ourselves, others and the objects in our surroundings. The physical boundary between our bodies and the outside world could define our sense of self, although that would ultimately reduce our subjectivity to biological anatomy. More precisely, then, we must come into being as subjects when we see, when we take objects in our field of sight or any of the five senses. Subjectivity is brought into being through sensory experience that brings objects into, for example, a subject's field of vision. This, most basic interaction with the world around us, then provides us with the raw experiential material for thought processing in the mind.

Subjectivity, then, precedes any ideas about what we each refer to as the 'I' or ego, when we use language to make claims about ourselves and the world around us. Our subjectivity or sense of self is, therefore, the first boundary between ourselves and the world. Without this subject-object dichotomy, without the distinction between self and world, without that initial break between observer and observable, there would be no self to talk about. Put directly, this gap between the self as subject and all the objects in the totality of the Universe is the first characteristic that we all share as human beings. It enables us all to tell the difference between each of us and the outside world. Human subjectivity or selfhood, however, is not by itself a guarantor of human thought, as we shall see.

The first modern philosopher of the thinking subject or human mind was the seventeenth century Frenchman Rene Descartes (1596-1650). His oft quoted statement, "I think, therefore, I am", from his book *The Meditations* (1641), was the result of a thought-experiment, when he decided to bring into doubt the reliability of all his sense experience (such as seeing and hearing). From the position of believing his whole life could be a dream, rather than reality, he arrived at the conclusion that the only aspect of his life he could not doubt was his thinking. For Descartes, thinking proved existence or being. The subject, then, lays the foundation for the modern Cartesian notion of the mind. Descartes claimed to prove his existence through thinking, but the subject who senses (even though Descartes subjected his sense experience to radical doubt) laid the foundation for him to reach this conclusion. Who or what sees the world prior to any thought? Sight precedes thought; however much we may doubt it, we see before we think.

The notion of selfhood, then, the subject-matter of this book, is in line with the Cartesian scientific observer, whose legacy remains very much alive today in the natural sciences, especially physics, where observations of the large-scale Universe (stars, planets and galaxies) and the small-scale (atoms, electrons and quarks) are based on the subject-object relationship between the observer and the world. The success of modern-day science is due to the disciplined application of physical observations made by humans within a real world, in order to find out the truths about the laws of nature and how everything has physical cause and effect.

Earlier, we posed the question of whether machines can think. As such, we must first examine the idea of the machine subjectivity. This can all be encapsulated with another question: do machines see? Do the most advanced robots in the world have the ability to see the world in the same way as humans do? If robots can see, then according to our definition of the subject who senses, they could be said to have the basic ability to have selves and think. If robots have subjectivity, then, how are they any different to us?

This book presents a theoretical basis for machine subjectivity. Some might argue against this, claiming that thought separates us from machines, that the human capacity for thinking is what makes us different. They claim that only we humans think and that robots do not; we are conscious beings, robots are not; we are alive, robots are not. These arguments are addressed in this book, because sensory experience, primarily sight, has been identified as laying the foundation for the thinking mind. What we are also seeking here is the understanding of ourselves.

This is not based on whether we can think or are conscious, but on if we *and* machines can sense the real world, whether we can both see and touch it. This will guide us in any further inquiry into thought and consciousness. It matters more whether machines can see reality than whether they think about it, because it is our senses that connect us directly to the world.

That today, the self is simply the free autonomous and rational being of the seventeenth century is heavily criticised by academics and intellectuals, particularly in the humanities. This rational subject (one who reasons) is the product of the scientific revolution of that era, that spanned from the sixteenth to eighteenth centuries. It was during this time that thinkers like Newton, Descartes and Galileo installed the subject of scientific rationality. The developments of the nineteenth and twentieth centuries, particularly in the field of psychology, with Carl Jung

(1875-1961) and Sigmund Freud (1865-1939), displaced the now apparently simplistic model of the rational subject. According to the psychologists, humans are not simply rational beings, as science would have had it. Rather humans are more like 'rational animals', a notion that originated from the philosophers of Ancient Greece. In other words, in addition to our ability to reason, we are beings who have instincts, drives and impulses.

In short, the rational subject of the eighteenth century lived as an automaton in the clockwork Universe of the great English physicist Sir Isaac Newton. This almost automaton subject could be reduced to his bodily mechanism, to cogs and bolts. In other words, the rational man could reduce himself to his mechanical functions. Now that, in the twenty-first century, we are on the brink of seeing the emergence of the robot, an autonomous, rational machine, one who is seemingly taking the place of the modern man (of science), us humans are left clinging with our fingernails to the last residue of our common experience: emotionality. The problem is that as we are becoming more confined to the internet, our social relations embedded in the electronic dimension, we are being gradually displaced by robots in the real world. Moreover, this mechanised revolution is taking place right under our noses in the automation of work.

Robots have been replacing humans in the workplace for some time and this is a growing cause for concern. According to the Bank of England in 2015, up to 15 million jobs, which represents 44 per cent of UK employment, are at risk of automation. Over the next 20 years, the forecast is that robots will replace humans in these 'at risk' jobs. The danger is particularly posed towards the service sector that the UK and other post-industrial economies depend on. The same figure for the USA is 80 million jobs under threat of mechanisation. The replacing of service sector jobs has been taking place for some time now, self-checkout tills at supermarkets being the clearest, everyday example. More significantly today, robots may have the capacity to replace humans in jobs that require creative thinking. Tragically, machines are making many humans jobless and due to today's imperative of technological efficacy, human labour is becoming expendable. Another example is the Uber taxi, with satellite navigation (satnav) technology, replacing the traditional London black cab, where drivers have to memorise the street locations, according to a test called The Knowledge.

Machines Who Think: The Rise of Artificial Intelligence

What is Artificial Intelligence? Those who strongly support Artificial Intelligence (AI) say the following: humans will one day create a robot as intelligent, smart and conscious as we are. The belief in this eventuality and its possible consequences, often disastrous and dystopian, has been sketched out in many books and films. This started in 1920, when the Czech playwright Karel Capek first coined the term 'robot' in his play *Rossum's Universal Robots*. Here, the story went, a robot-led rebellion causes the extinction of the human race. Later, after robots were properly developed, Hollywood blockbusters, like *The Terminator* (1984) and *Robocop* (1987), presented us with more negative dystopia about the rise of AI in the future.

The theme of all these movies is that somehow humans have a capacity to create intelligent mechanical beings outside of ourselves. This vision of the future seems to generate dystopian fantasy, exaggerating the risk of robots taking over and running terribly out of control. It was Mary Shelley who got there first, however. Her classic novel *Frankenstein* (1818) was written in an age of heightened scientific, technological and industrial change, though rather than resort to dystopia, Shelley's vision criticised the role of man in making machines.

Today's AI is like a belief-system that encases much of the thinking that goes into our approach to robotics. The idea, of course, is that the machines being developed today are increasing so much in sophistication that they have a form of humanly created intelligence. As with any belief-system, it cannot be established reasonably when or how robots will achieve human-level intelligence, but the ambition is there amongst the AI scientists and tech gurus of our time. The idea of AI then is challengeable and the controversies that surround it will remain, such that we must now no longer postpone questioning subjectivity and issues concerning the human self, machines and who we (and they) really are.

In the absence of a generally applicable scientific definition of what intelligence is, any attempt to place it within a research framework of AI risks a descent into anthropocentricism: in other words, projecting human characteristics onto machines. Rather than understanding what a robot is, all AI is doing is imagining that one day human subjectivity can be reproduced in inanimate objects. This skips the ongoing questions we must ask of ourselves concerning how little we know of what it is to be human. AI, then, needs to be rigorously re-examined; like pre-Copernican thinking that placed the Earth at the centre.

The best argument against AI was actually made over 300 years ago. It remains unrefuted and warrants close examination today. In 1714, the German philosopher Gottfried Leibniz outlined the central disagreement we have with AI. In Part 17 of *The Monadology*, Leibniz compared the human mind to the ultimate potential of machines. In *The Mill Argument*, he stated the following:

> If we imagine a machine whose structure makes it think, sense, and have perceptions, we could conceive it enlarged... so that we could enter into it, as one enters a mill. When inspecting its interior, we will find only parts that push one another; we'll never find anything to explain a perception.

Leibniz says that if we are to seek perceptive intelligence in a machine, it cannot be found in the human assembly of it, however large and complex it might be. It is not the composition, complexity or structure of a machine, such that one day you will suddenly find consciousness, perception or intelligence in it. Leibniz pointed to this simple fact in the eighteenth century and today it still needs to be addressed.

Even if you build a hypercomputer the size of a city, with all of its working parts operating like an enormous brain, when we enter this machine, we could see all its parts, all the circuitry, all the 'cogs and wheels' that make it function. In this sense, we could never see anything in the vast machine that possesses intelligence. With Leibniz, we see that AI today is a 'paper tiger' argument that does not even stand up to eighteenth century philosophical scrutiny. However complex and advanced we might think we are, having already made basic robots, in the end there is nothing we can add to the material composition of the machine to give it the ability to perceive intelligently. Again, any new argument for AI must first address subjectivity and the definition of perceptive awareness.

Leibniz has not stopped AI from penetrating our collective psyche through mass media. Every day we are presented with apocalyptic scare-stories about the future. In 2014, when interviewed by the BBC, Professor Stephen Hawking, physicist, warned that: "The development of full artificial intelligence could spell the end of the human race." Technology billionaire Elon Musk has worryingly claimed that AI is 'our biggest existential threat'. Even Microsoft billionaire Bill Gates, himself responsible for the most significant shift in the development of the desktop computer, has described AI as a 'concern'.

These scare-stories do not take into account the reality. Our misuse of too much technology is a threat to our existence, not because of some far-flung doomsday scenario, projected by the industrial and academic proponents of AI, but due to a lack of recognition of how too much technology in our everyday lives, the lack of face-to-face social interaction, creates distance in human relations and separates us from the real world. If technology is taking the place of religion, then AI is a worldview that is framing our response to the rapid changes that the promise of electronica has offered over the last 50 years. Rather than creating new social freedoms as envisioned by the captains of the silicon industry, we find ourselves deeper and deeper in the quagmire of technomania, an extreme enthusiasm for technological novelty in place of normal human relations that some would reluctantly consider a thing of yesteryear (even if they are accused of technophobia, the fear of technological advance).

The Singularity: A Point of No Return

The term Singularity in relation to technological change can be traced back to a conversation in May 1958 between two celebrated mathematicians, Stanislaw Ulam and John von Neuman. Giving an account of this, Ulam later wrote:

> One conversation centered on the ever accelerating progress of technology and changes in the mode of human life, which gives the appearance of approaching some essential *singularity* in the history of the race, beyond which human affairs, as we know them, could not continue.

Whatever we may think about the ever-increasing rate of technological change and the inability of humans to cope with the accelerative expansion of the digital world, there is no reason for us to predict a dystopian point in the future, today popularly termed 'the Singularity', when the machines take over. It is, however, entirely feasible (as Ulam and Von Neuman did) to discuss the negative impact of technology on human relations, since – as they predicted – technology is today inserting itself rapidly into human affairs.

Nowadays, the term 'Singularity' has been hijacked to describe imminent catastrophe due to AI-takeover. In a 2005 book, *The Singularity is Near*, so-called 'futurist' author Ray Kurzweil describes a near future when computer processing ability will have increased to such an extent that it will be impossible to distinguish between machine and human intelligence: 'The Singularity will allow us to transcend these limitations of our biological bodies and brains ... There will be no distinction, post-Singularity, between human and machine.'

Kurzweil states that this will occur in 2045, backing it up with a calculation from a generalisation of Moore's law (a theory posited in 1975 by Gordon Moore, co-founder of Intel, who suggested that computing power doubles every 2 years). For this 'calculation', Kurzweil also confines human brainpower to computational processing ability. In other words, he suggests that the human mind is reducible to calculations per second (cps).

This calculation is based on the brain's billions of neurons, their thousands of connections to other neurons, with each connection performing hundreds of calculations per second. Absurd as it might seem, Kurzweil claims the human mind can be put on an equal level with a computer, reducible to processing ability. Quite simply, according to Kurzweil, whereas computers are based on transistors, the mind operates via neurons; otherwise there's no real difference. Following this logic, Kurzweil goes on to say:

So we will be producing about 10^{26} to 10^{29} cps of nonbiological computation per year in the early 2030s. This is roughly equal to our estimate for the capacity of all living biological human intelligence...This state of computation in the early 2030s will not represent the Singularity, because it does not yet correspond to a profound expansion of our intelligence.

Kurzweil goes on to make a prediction that by the mid-2040s 'the [computer] intelligence created per year...will be about one billion times more powerful than all human intelligence today'. This unfounded claim has not stopped Kurzweil from working for Google and receiving The National Medal of Technology (1999) from the highest office in the United States, the Presidency.

What is the problem with the Singularity? Comparing human intelligence to a machine that performs calculations is simplistic and not scientifically proven. Calculation is one aspect of human intelligence, one that is needed for playing chess, for instance. If calculation is your only concern, then computers have surpassed us in that field, as Deep Fritz proved against our grandmaster, Vladimir Kramnik. In addition, the multifarious, indeterminate connections in the neuronal system of the brain cannot yet be compared to the circuitry of digital computers that humans develop themselves. Human intelligence may not even be limited to the brain. Computers could just be an extension of our intelligence in the silicon realm. Is the smallest cellular biological unit of the brain, the neuron, the same as the computer transistor? Is the neuron a simple on/off device, like a transistor switch? Neuroscience suggests otherwise. We are only *beginning* to understand how the human brain works and observational scientific tools like fMRI (frequency-Magnetic Resonance Imaging) form the start of this exploration. If the brain is just like a computer, then how does it work?

Anyway, why does there have to be a technological singularity in the first place? If one day, machines are supposed to become as intelligent as we are, at what point did their intelligence bifurcate or separate from ours in the first place? In other words, if we are to project some point in time that machines will become as intelligent as us, when was it that they become intelligent in their own right to begin with? Again, if you cannot refute Leibniz, there is no need to imagine that the so-called 'Singularity' is near.

From the time of Charles Babbage (1791-1871), who invented the first 'calculating engine' in the nineteenth century, computers themselves have remained also the product of human intelligence and creativity, not only of human calculative ability. The intellectual process that goes into performing a calculation is distinct to that which goes into invention and creativity. If we define intelligence as the ability to see both sides of an argument or comprehend, learn and dissect every aspect of a question or problem, then machine intelligence remains radically different.

The idea that one day, machines will surpass us in the same capability that we needed in their development, lacks philosophical rigour, because it disregards any concept of subjectivity, the self and the mind. If we are to reduce thought to calculation, then of course there will arrive a day when the sum total of the chips and wiring in the world will outdo humans, in what machines were intended to do in the first place: that is, calculate.

Today, the point of the technological Singularity could serve one purpose only: to explain, without rational justification, the anxiety of why the development of technology could lead to the end of humanity. This threat of technihilation lacks a rational basis, as opponents of Kurzweil have stated. In support of the anti-Singularity position, the Harvard University cognitive scientist Steven Pinker said in 2008 that the Singularity will 'never' occur, adding, 'There is not the slightest reason to believe in a coming singularity.'

Furthermore, as the virtual reality pioneer Jaron Zepel Lanier explains, there may be a hidden agenda behind these narratives, in devaluing human agency in favour of technological determinism:

The reason to believe in human agency over technological determinism is that you can then have an economy where people earn their own way and invent their own lives. If you structure a society on not emphasizing individual human agency, it's the same thing operationally as denying people clout, dignity, and self-determination...to embrace [the Singularity] would be a celebration of bad data and bad politics.

The significance of the Singularity is that it frames our understanding of many of the developments we see in science and technology today. These also seep into the popular consciousness in Hollywood films such as *The Matrix* (1999) and confine our way of thinking about machines along the lines of whether they are beneficial *for us*. Of course, most of these movies are made using the same state-of-the-art technologies they seek to undermine. If you are worried about the machines taking over, why watch some hypothetical multi-million dollar budget movie called *Robocalypse Now!* made with high-level computer graphics? To escape the Matrix, why immerse yourself in a larger cinematic one?

In reality, anyone who has encountered a robot will know that they can barely handle stairs, let alone plot the destruction of all life on Earth as we know it. Then why is it that in 2014 Steven Hawking, one of the most eminent minds on the planet, said: 'AI could spell the end of the human race.'? The lack of rational justification for these statements is all-too-common and plays on our irrational fears, our lack of understanding of the relationship between man and machine and above all, a desire to create an imaginary threat where there is none.

There is no such thing as *The Singularity*, beyond an aside made in a conversation between two mathematicians in 1958, about how human relations could be affected by technology. The danger we are faced with today emanates purely from ourselves and our thoughtless use of technology. A destructive nexus has appeared, where a rapid expansion in technology designed to facilitate our thought processes has been met with selfish disregard for others. We must examine the consequences of our digital sedation, our disengagement from one another and our detachment from ourselves.

Press Ctrl-Alt-Del to Log-out and Drop out!

The Structure of *The Universal Subject of Our Time*

This is a book that aims to outline the scientific and philosophical background that has shaped our present-day understanding of subjectivity, in the midst of unprecedented technological change and on the verge of potentially major breakthroughs in robotics and machine intelligence. There are two sides to every coin and technology has its benefits and drawbacks. Its misuse partly motivates *The Universal Subject of Our Time.* This book also serves as a guide to the world we inhabit today. In the wake of what is presented here as a breakdown in human relations, due to an over-reliance on technology, *The Universal Subject* also seeks to shed light on the underlying reasons why human and machine subjectivities are intertwined and often seem in conflict with one another.

In the first chapter, we will look at what is meant by subjectivity today, with reference to scientific and philosophical thought. The notion of the subject is one that we take for granted. A brief study of the most significant philosophical trends of the late twentieth and early twenty-first centuries indicates that subjectivity has reached a dead-end. What is termed in this chapter as *The End of the Self,* is the result of developments in post-modern thought, a philosophy that has dominated the academic world, particularly in the humanities, and represents an unbridgeable rupture with any fixed notion of the subject or self. What post-modernity promotes is a scepticism towards any so-called 'grand narratives', any fixed, objective truth, even concerning the existence of the subject. In many ways, the post-modern narrative taps into a human yearning for absolute freedom of being. Rather than identifying the subject as a self-subsistent entity, it draws on concepts such as power relations (Foucault), difference (Derrida) and ideology (Althusser) that bring our sense of subjectivity into being. It is a characteristic of these times that, especially in popular culture, the subject or self can not only be reinvented, but has no fixed meaning.

The second chapter considers how the natural sciences challenged post-modernity at the end of the twentieth century. The sciences, particularly physics, hold true to the notion of the self, derived from the scientific revolution of the eighteenth century, when astronomical observations of the solar system prevailed against religious doctrine about the stars and planets. Even today, scientists support the notion of the rational and observant subject at the centre of their work, even if they may have lost the debate to post-modernism. On the other hand, for post-modernists, nothing is objectively true and, *in extremis*, nothing is subjectively true either. Although science continues to adhere to the principal of objective truth and also affirms the subject of observation, the so-called era of 'post-truth' we are said to inhabit today suggests that post-modernism is alive and well. Did post-modernism, with its radical denial of fixed truths, suggesting that science is just another human construct, really win the debate? It does seem that scientific progress is not guaranteed. In this chapter, we draw out the lessons and legacy of this intellectual show-down.

Part II of this book is an outline of how subjectivity is resurrected as a result of machine subjectivity that goes beyond our preconceived ideas about the centrality of the human self. As such, in the third chapter, an account is given that takes the subject of the machine as the starting point for a new approach to understanding not just robot (or mechan) intelligence, but ourselves. In this book, the term *mechan* (equivalent to human) is introduced and means intelligent machine or robot machines have a long history, concurrent with the evolution of human beings. We are a tool-making species from our very beginnings. A machine is a complex tool and technology has resulted from the evolution of primitive tool-making. When we first created fire, foraged for food or hunted animals, we used tools that separated us from nature. These material tools, simple implements to begin with, followed a parallel course in evolution, perhaps even supplanting our biological evolution, towards what we see as today's technology. By inserting themselves into the gap between us and nature, machines are developing into entities that challenge who we are today. Machine subjectivity is, therefore, directly bound to the material Universe and its properties. Here then, we also examine the fundamental nature of the cosmos, particularly black holes or – as we call them – black stars, to see how they may generate subjectivity (black stars defy the subject-object dichotomy in that they cannot be seen). In this chapter we also explore the implications of this theory and question whether our over-reliance on the mechanical thought process has signalled human self-defeat.

Then, in the fourth chapter, we embark on an appraisal of the human subjectivity, in light of machine subjectivity. The proposal here is that the human subject is mediated by technology and, therefore, primarily defined in its relation to technology and machines. If there is anything, however, that makes us distinct from robots, it is our technologically mediated selfhood. We take our subjectivity for granted, something that we are almost born with, a guarantee of who we are. What particular aspect of who we are defines our sense of being? If the human body anatomically presents us with everything you need to be who you are then where is your self then situated? What if technology is desensitising us from the real world? Are we observers of a digital world that is radically severed from our physical being? We then draw out the implications of the position taken here that the human self is mediated by technology in relation to the physical world around us.

In the fifth chapter, our discussion of the self ends with the development of a future course of history that pits man against machine. The more we bind machines to us, the more they take on our characteristics. The more dependent we become on technology, the more technology takes over our thought processes. The more we use technology to attain a sense of mastery over the world, the more consumed we become by it. This dialectical movement, man vs machine, this ongoing struggle for mastery, is taking place not on the pages of science-fiction novels or films, but every day. The end-goal is the Universal Subject or a selfhood that relates to the Universe as it is. The overuse of technology in our lives now leaves you with a choice. The implications are examined: with reality displaced in the pursuit of apparent technological freedoms, we are instead seeking to divest ourselves from the often burdensome sense of self and entering technological nihilism in the process.

In the final, conclusory chapter, we will take a look at how the human mind and that of the machine can be consolidated via a new reality, following a virtual fissure that has emerged in the wake of this technological disjuncture. Having established that our sense of selfhood is reliant on technology, we must identify how the human mind, now inhabiting the virtual, reconnects to the physically real realm of the body. By examining how our bodies respond to external stimuli and locating precisely where this connection between mind and body takes place, the purpose of this chapter is to reinstate the link between our minds and our bodies. Having introduced machine selfhood at the centre of our understanding of ourselves, we need to reconnect our minds to our bodies, due to the emergence of virtual reality (VR). *Our New Reality* is needed to show that we cannot exist as disconnected selves floating around a world of virtual cyberspace, because we are human selves fully connected to our real natural environment. There also is, at the end of the book, a series of aphorisms that can help inform the reader about the motivation and purpose of the Universal Subject. Twenty Steps to the Universal is a series of emotively written paragraphs that outline how we came to find ourselves in this technological situation.

Annihilation by technology is the conversion of human thought into packets of machine-processed data without recourse to human interaction. Rather than freeing your mind to pursue creativity and meaningful relationships towards one another, the double-edged sword of technology has become the primary means of escape, from each other and, above all, our subjectivity. *The Universal Subject of Our Time* serves as a diagnosis, remedy and, ultimately, optimistic palliative for our recurring modern ailments in the best and worst of times we are living in.

PART I

Chapter One

The End of the Self

But what then am I? A thing that thinks. What is that? A thing
that doubts, understands, affirms denies, is willing, is unwilling,
and also imagines and has sensory perceptions.
Rene Descartes, The Meditations

'O my friends, there is no friend!', used to proclaim the
influential post-modern French philosopher Jacques Derrida at
the beginning of his lectures in the 1970s. The statement sought
to undermine preconceived notions of, not only friendship, but
of subjectivity. For, what is a friend? Another self, outside of
one's own, with whom you share common characteristics and
interests? Then how can one objectively identify friendship,
when a friend is someone whose selfhood is most closely aligned
to one's own? The notion of the friend as an Other, appearing
separate, yet of one's own self, seems contradictory. Derrida's
intention, as his work demonstrated, was to question any notion
of subjectivity that we may take for granted. The idea of the
friend provided Derrida, at least, with a starting point to launch
his critical project against traditional ideas about subjectivity
inherited from the Enlightenment.

The aim of this chapter is to show how early modern
ideas about the subject, as a free, self-subsistent and essential
individual, were undermined by psychoanalysis and finally put
to the test by post-modernism and post-humanism.

The Modern Subject: from Descartes to Kant

During the European Enlightenment, an era said to span from
the sixteenth to eighteenth centuries, religion was sidelined
and, in its place, science took centre stage. No longer was the

individual subject constituted by a theological framework; modernity ensured that the human subject was unbounded by such concerns and that ultimately notions of scientific progress and advance began to supersede traditional doctrinal arguments derived from religion. To this day, the legacy of these developments is most tangibly felt with progress in science and technology.

Rene Descartes (1596-1650) became the leading modern philosopher to outline a theory of the subject with his dictum 'I think, therefore, I am.' The statement remains alive in the popular consciousness today, even making its way into Hollywood movies about intelligent robots such as *Blade Runner* (1982). The Cartesian subject was born in a fluctuating field of radical doubt: in formulating his theory of the subject, Descartes brought into question his very existence and all of his sense experience. Finally, in his book *The Meditations*, he came to rest on the certainty of his thinking activity as the guarantee of his existence. Following Descartes and the Enlightenment, the human subject became synonymous with a free, rational and self-sufficiently unitary being or individual. The only criterion for subjectivity became the thought process itself, the ability of humans to process sense experience and memories rationally. The activity of the rational mind itself was enough to prove this.

In the late eighteenth century, Immanuel Kant (1724-1804) took up Descartes' mantle as the most significant philosopher of subjectivity with his work *The Critique of Pure Reason* (1781). Here, Kant outlined his view that simple sensory perception is not enough to explain the workings of the human mind and that the mind itself, in its make-up and constitution, influences our perception of the world around us. According to Kant, the human mind shapes the world around it, giving rise to knowledge. Rather than being an empty vessel into which sense experience is poured, the mind itself is self-conscious; there is a self that conditions what we see, hear, taste, touch and smell.

The Enlightenment is said to have been brought to a conclusion with the far-reaching events of the French Revolution (1789) and, as such, it was Immanuel Kant's work that marked its philosophical denouement. The legacy of the Enlightenment remains with us today in notions of the individual, freedoms and progress, but most significantly for our discussion of subjectivity, it is in the scientific process where its influence has survived and remains entrenched.

The Subject in Psychoanalysis

The nineteenth century, following the political turmoil of the French Revolution, saw the Enlightenment being brought to a close. In place of older ideas about subjectivity, emerged new concepts that tried to explain the apparent potential for irrationality and unpredictability in human behaviour. Were there hidden, unconsciousness drives and desires that motivated the self, outside of what is readily apparent? The emergence of psychoanalysis then, in the late nineteenth century, was the result of an era that was challenged not only by revolution, but by an awareness that human behaviour tended not always to have an obvious cause. Impulses emerging from an underlying seething unconscious seemed to have at least an equal influence on human behaviour and even history.

The unconscious psyche or mind formed the starting point for Sigmund Freud's conception of human subjectivity. Whereas the rationalists of the Enlightenment saw the mind in almost clockwork terms, Freud saw the human subject as one who treads a fine line between an apparently rational conscious awareness and a deep-seated unconscious. Dreams were the most direct evidence, for Freud, about the unconscious, for it was during the dreaming process that man's conscious defences are let down and the unconscious has almost free rein to express itself.

Although Freud's work challenged traditional notions about

the human condition, what was left almost wholly intact was the fixed notion of subjectivity. From Descartes through to Freud, there is no question of the intactness of the human subject as a singular whole, one who observes, has memories and is entirely self-sufficient. What Freud showed was that the subject may not be necessarily governed by rationality and could instead be influenced by unconscious irrationality. Subjectivity, therefore, had not yet been uprooted by Freud, but the appearance of the unconscious undermined previous conceptions of the self as an independent product of self-volition and rationality.

Enter the Post-Modernists

By the latter half of the twentieth century, following the economic turmoil and disastrous wars of the earlier years, the emergence of a new school of philosophy, called post-modernism, challenged the very foundation of subjectivity. Post-modernism, with its critique of fixed objective truths and so-called 'grand narratives', saw subjectivity as being a target for critical thought. For postmodernism the notion of a singular objective truth was rejected in favour of scepticism towards all truth claims, especially those associated with the grand narratives of, for example, the truthseeking scientific endeavour or sweeping accounts of history. Not only, however, were truth claims subjective, but the very basis for giving a subjective account of the world had to be scrutinised, as subjectivity itself was seen as the grandest of all narratives.

Thus, any narrative of subjectivity, especially emanating from the Enlightenment, underwent philosophical critique by the post-modernists. The very notion of the self-subsistent subject, the human self, the idea that there is an irreducible subjective core to our being, was picked apart by a series of continental philosophers, most notably Michel Foucault, Jacques Derrida, Louis Althusser, Jacques Lacan and Jean Baudrillard. By the end of their interventions, the banner of human subjectivity, raised wondrously aloft by Enlightenment figures such as Descartes and Kant, had been shredded to rags (at least in the humanities). Here we will examine each of the post-modernists in turn:

1. Michel Foucault (1926-1984) and Power

Foucault said that the subject is formed merely through power relations. Power forms the subject. In the production of subjectivity, says Foucault, it is the internalisation of that discourse of power that creates the subject. Foucault gives an alarming account of subject formation in his books *The Birth of the Clinic* (1963) and *Discipline and Punish* (1975). Here and elsewhere, he outlines that our social environments are bounded by institutions such as law enforcement and the psychiatric profession that confine us into being who we are. We exist as subjects only in so far as we are moulded by the state intuitions that draw lines around who we can potentially be. Subjectivity is no longer essential; it is secondary to power. For Foucault, power relations are non-subjective and do not necessarily derive from the choices or decisions of individual subjects. Subjects do not come into being naturally or on their own, but are rather products of the instruments of state power. The human self had been subordinated to power, in Foucault's eyes.

2. Jacques Derrida (1930-2004) and Difference

Whereas for Foucault power precedes subject formation, for Derrida it is the notion of *difference* that takes priority. Difference does not equate to the difference between individual subjects, such as you, the reader, and I, the author, but in terms of *what differs* in order for there to be subjectivity. There must be difference first, in order for the subject itself to form. The primordiality of difference calls into question the subject itself. In order for us to posit a subject, any subject, there must first be difference. For Derrida, then, the subject itself can be deconstructed into its differing constituents. Taking the original subject-object dichotomy, for example, deconstruction would question whether it is primordial at all. Difference would have it that perhaps the two terms 'subject' and 'object' are mutually consistent and that one would have to go further to identify what truly differs in order for this dichotomy to come about. In this way, subjectivity, as a product of difference, is subordinated.

3. Louis Althusser (1918-1990) and Ideology

Similarly to Derrida and Foucault, Althusser was concerned with demoting the position of the subject. Here, ideology has the property of constituting individuals as subjects, such that subjectivity is secondary to ideology. What is meant by ideology is the body of ideas fostered by a society that conditions how subjects within that social order perceive the world. Althusser was concerned particularly with state ideology, saying, "Ideology interpolates individuals as Subjects... ideology has the function of constituting concrete individuals as subjects." In other words, subjectivity is the mechanism by which ideology conditions the members of any society to think, speak, act and feel. It is the recognition that you receive as an individual person, be it in the electoral voting process or any other state mechanism that makes you into a subject. There is no such thing as innate subjectivity. Perhaps similarly to Foucault, Althusser viewed subjectivity as having an oppressive dimension: we do not choose to be subjects, it is foisted upon us by state ideology that is concerned with the maintenance of any society. In this sense, subjectivity is a socially-constructed phenomenon that has no fundamental or necessary existence outside the auspices of a society that ideologically constitutes itself.

4. Jacques Lacan (1901-1981) and Language

The subject according to Jacques Lacan is very much born into language. Prior to language, prior to the ability to symbolically attach signifiers to images around us, the subject in us does not even really exist. If language is absent then how can there be subjectivity, since without any language, there would be no way of defining the difference between yourself and another person. Language is what constitutes us as distinct subjectivities.

5. Jean Baudrillard (1929-2007) and the Simulacra

For Jean Baudrillard subjectivity is also preceded and subordinated by the idea of the simulacrum. In his most influential work, *Simulacra and Simulation*, (1981), Baudrillard argues that all modern-day human experience is a simulation of reality composed of signs and symbols (or *simulacra*). In other words, there is no objective reality and furthermore, if there is subjectivity at all, it is constituted by the simulation that it inhabits. All-in-all, Baudrillard's critique of the real world we are presented with retroactively demolishes any notion of subjectivity we may seek to anchor our sense of reality in. Unsurprisingly, his ideas inspired the Hollywood sci-fi film series of *The Matrix* (1999).

By themselves, each one of these five post-modern philosophers could challenge Enlightenment subjectivity. Together, however, with the support of other thinkers of their ilk, the notion of subjectivity that had been inherited from modernity was smashed asunder. Power, difference, ideology, signs and language were all of greater significance in shaping who we are than the fixed, immovable, self-sufficient subject of Descartes. The subject is a product of power relations, difference, state ideology or just meaningless outside of language. These perspectives were enough to relegate the importance of subjectivity as a serious focus for debate in the last decade. Subjectivity is of secondary concern. The self is constituted purely through human mechanisms such as the state, for example, which is concerned solely in preserving itself. Descartes' *thinking self* had been dethroned entirely and replaced with an array of new post-modern ideas.

Subjectivity, Technology and the Rise of the Cyborg

In the wake of the destructive levelling of the terrain of subjectivity by post-modernism, new ideas about human subjectivity began to emerge. Most notably came the idea that human subjectivity had no fixed existence and that the emergence of technology suggests that mankind may be on a *post-human* path. The cyborg, as we will see, is what could emerge with the end of subjectivity.

In her ground-breaking essay *A Cyborg Manifesto* (1984), Donna Haraway suggests that we have already become a mixture of human and machine. The cyborg is part-machine and part organic human, an integration of nature and technology. She states in the essay:

> By the late twentieth century, our time, a mythic time, we are all chimeras, theorized and fabricated hybrids of machine and organism; in short, we are cyborgs. The cyborg is our ontology; it gives us our politics. The cyborg is a condensed image of both imagination and material reality, the two joined centres structuring any possibility of historical transformation.

In other words, the culmination of the post-modern dismantling of subjectivity leaves us at the end-point of the cyborg. We are all cyborgs, already.

According to Haraway, the traditional boundaries between human and machine have broken down. We are becoming more like the hybrid products of technology, whilst at the same time robotics is producing machines that are becoming more like us. Science has failed to discover something essentially different in human nature that could make us distinct from intelligent robots. She later explains that whereas the technology and machines we use are becoming increasingly 'lively', humans themselves are becoming relatively 'inert'. The integrative point, therefore, at the end of this hybridisation process, is the half-man, half-machine cyborg.

If there is no such thing as human subjectivity, if nothing in our essential nature can be pinpointed that differentiates us from machines, then we slip into cyborgism. Of course, many will argue that creativity and emotion are the key factors that make us human; but even with all the advances in neuroscience, such as brain-scanning by frequency-Magnetic Resonance Imaging (fMRI), we have been unable to identify a single human thought process that cannot be replicated by a machine. It is not enough to state that machines cannot be creative, when we are unable to scientifically prove what human creativity is. We may claim to be creative, but the ultimate pinnacle of our creative energies today are invested in producing the likeness of our very selves: the intelligent robot.

The Emergence of Post-Humanism

The central strain of thinking to emerge then at the end of subjectivity is post-humanism. Perhaps the first exponent of post-humanism in science-fiction was the visionary British author Olaf Stapledon (1886-1950). In his book *Last and First Men: A Story of the Near and Far Future* (1930), Stapledon outlines the future course of humanity on an unprecedented scale. From the present, going forward up to 2 billion years, Stapledon identifies the emergence of 18 new human species, of which our own is merely the first.

Along the course of this sequential development of man to the post-human, perhaps the most significant of Stapledon's science-fictional leaps is the fifth man. Stapledon describes the fifth man as an artificial human species designed by the previous (fourth man) that are on average 'twice as tall as the First Men, and much taller than the Second Men'. After eliminating the fourth men, the fifth men develop an unprecedented level of technology, even enabling man to make Venus habitable.

Stapledon's artificial human species foresaw Haraway's cyborg. At the root of post-human theories is, of course, the end of subjectivity. It is not that we cannot speak of subjectivity when the cyborg has emerged, but that our traditional concept of the human subject, a product of the Enlightenment, must be reconsidered. The question of whether cyborgs have selves can be approached; however, Haraway in her writing makes it clear that the defining feature of the cyborg is its post-human absence of subjectivity. There is nothing that can subjectively constitute Haraway's cyborg; the cyborg is beyond even the human conceptions of power, ideology, language, signs and difference of our post-modernists.

In this way, technology, according to Haraway and other post-humanists, nullifies subjectivity. Today, we already live in a world enmeshed with high-tech gadgets that we use every single day in our lives. Is the human talking and networking on a smartphone not the first step in the emergence of a human-technological cyborg? Has the cyborg already arrived in our lives and are we already partaking in a new cybernetic culture with our use of technology? Are the apparent freedoms that technology gives us not the driving factor in a new post-humanist digital revolution, which we are unconsciously demanding? The seemingly untrammelled desire that we are displaying for new technologies, especially in the arena of communication and social networking, suggests that the post-human drive is with us already. We are already seeking to break out of the limited boundaries that we created for ourselves in the humanist outlook.

Haraway, the matriarch of today's generation of post-humanists, set out the end-goal of human evolution as being the cyborg in 1984. Another post-humanist, N Katherine Hayles, with her 1999 book *How We Became Posthuman*, brought the theoretical basis of post-humanism to widespread attention. In 2017, Stefan Lorenz Sorgner and James Hughes, with Penn State University Press, established *The Journal of Post-human Studies* in order to place the field of post-humanism on an academic footing, such that present questions on the issue could be taken on. Finally, this year, the publication of Yual Noah Harari's *Homo Deus*, a bestselling book that predicts a post-human future culminating in a supernatural 'man god', has popularised the idea that there could be a state being beyond the human. All-in-all, it has become clear that in recent years the transformative impact of new technologies has pushed human thought out of a comfort zone about the future. We have become overawed by the products of technological imagination, not only in terms of our portable devices, but also in cinema, where visions of the future often feature man and machine either synthesised or in conflict.

The Limits of Post-Humanism

Do cyborgs think and, if so, about what? One of the ramifications of post-human thought is that, as cyborgs for instance, it becomes less and less possible for us to conceive of anything outside of our selves. If we are all becoming cyborgs what exists outside of ourselves? What is absent in the account given by post-humanists about our condition is a theory of subjectivity. The consequence of this is that humans begin to inhabit a vacuum, since it is only through subjectivity or sense of self that we can constitute our very selves as human.

As humans we can readily identify the characteristics that make us as such. Descartes pointed us towards thought. The cyborg, unless it is a state that we have already conceivably reached, is not one that can be immediately acknowledged. The cyborg, the actual physical enmeshment of man and machine, the cybernetic development of a hybrid, is to be seen as the end-point of our current development, but not one that can be easily defined.

Where post-humanism falls short, therefore, is that in positioning itself beyond the human, it radically abandons any mechanism for determining who we are. The Cartesian subject, as we have shown, exists subjectively through human thinking. The post-modernists undermined this and the post-humanists went further and rejected any conception of the human. In doing so, post-humanism has not addressed the central question of whether cyborgs can think, and if so, how does their thinking differ from ours? Post-humanism cannot answer this question and none of their formulations from Haraway's cyborg to Harari's *Homo Deus* address this central question.

It is thought and thought alone that determines human subjectivity from the Enlightenment onwards. Any attempt to derail this progression must present a theory to substitute this (such as the ones presented by the post-modernists), otherwise the entire post-human discourse falls irredeemably short. Not that we cannot appreciate the post-human science-fictional works of Stapledon, but in the absence of any theory it all remains just make-believe. However, even if post-humanism is yet to present us with a coherent theory concerning subjectivity and *thinking*, the work of the post-modernists Derrida et al (mentioned earlier) sustains and will continue to feed the work of the likes of Haraway and Harari.

Is It Really the End of Subjectivity?

In this chapter we have explored how the transition from the modern to the post-modern saw an evacuation of the subject. The subject itself, what we can describe as being the source of all subjectivity, was deemed unnecessary by social theorists, philosophers and psychoanalysts. The subject is expendable in favour of power, difference, language, signs and ideology.

As mentioned previously, the natural sciences remain the standard bearers of the Enlightenment theory of the subject. The observer of scientific exploration is one who works with objective truths. The subject of science seeks to establish the objectively existing laws that govern the Universe. Even in this age of quantum theory, multiverses and gravitational waves, the subject-object dichotomy in science remains intact. Since the subject as scientific observer forms its indubitable foundation, science was put on a direct collision course with post-modernism in the late twentieth century.

This clash of cultures, between science and post-modernism, was termed the *Science Wars* in some quarters. The next chapter explores the historical context of this development, its outcome and how, out of the mutually assured destruction and its fallout, a new paradigm could emerge that takes machine subjectivity or mechanical selfhood as its starting point. For, as science and the humanities locked horns in a debate over truth, high technology has crept up on us all and the robot may yet emerge as the final victor.

Chapter Two

Science against the Post-Modern

Postmodernism...has left behind a generation of academics in the humanities disabled by their distrust of the very idea of truth and their disrespect for evidence, settling for 'conversations' in which nobody is wrong and nothing can be confirmed.
DC Dennett, cognitive scientist

On 7 May 1959, the British scientist and novelist CP Snow (1905-1980) delivered a historic lecture in Cambridge that would identify a rift in the intellectual establishment that persists to this very day. Snow suggested, during the annual Rede Lecture, that a 'gulf of mutual incomprehension' exists between literary intellectuals, those from the humanities, and scientists. As a physical chemist himself, Snow bemoaned that the intellectual culture of 'the whole of Western society' had become apparently irreparably split and, furthermore, scientists had been given a lower intellectual status than their counterparts in the humanities.

In this chapter, we will explore how this fault-line developed into a fully-fledged conflict between science and the post-modernist humanities and what consequences the result of the ensuing conflagration had for our understanding of subjectivity. As we will see, the culmination of these so-called Science Wars is fast approaching now as both sides may yet be preparing for a final show-down over subjectivity.

From the Two Cultures to the Science Wars
During the lecture, Snow described how disappointed he was that in the company of 'highly educated' literary people, there was such a lack of understanding and appreciation of even the

most basic scientific concepts. At one previous gathering, he was provoked by a group of intellectuals who proclaimed 'the "illiteracy" of scientists'. In response Snow had asked them whether they could describe the Second Law of Thermodynamics. How surprising that they could not, suggested Snow. Yet, as he explained during the lecture, in referring to this physical principle, he was demonstrating the intellectual's scientific illiteracy by presenting them with the equivalent in science of the works of Shakespeare. It seemed that science and the humanities shared nothing in common. As Snow stated in 1959:

> The non-scientists have a rooted impression that the scientists are shallowly optimistic, unaware of man's condition. On the other hand, the scientists believe that literary intellectuals are totally lacking in foresight.

The lecture itself had a significant impact on both sides of the Atlantic, creating an arena for a debate that has lasted to the present day. Such was the import of the lecture that attempts have been made to bridge 'the gap of mutual incomprehension'. Most notable of these was a growing movement initiated by the literary publicist John Brockman, with his 1995 book *The Third Culture*, to create a new forum, where scientists could engage the public directly through the medium of popular science writing. The stated aim of the so-called 'Third Culture' was to empower scientists to take the place of the traditional intellectual in defining the questions that concern us most today.

Despite this attempt to bridge the gap between science and the humanities, the two-cultures divide, identified by Snow in 1959, had by the 1990s widened to such an extent that an intellectual show-down was inevitable. The terrain for this conflict became the philosophy of science, with frontline thinkers from the humanities, including leading post-modernists such as Jacques Derrida, taking on the physical realists, with the physicist Alan

Sokal leading the charge. At stake was not just scientific truth, the very principle of realism underpinning scientific endeavour, but – as we will see – the key issue of subjectivity.

For, if as the post-modernists contended, there is no objective truth and the subject itself is merely an Althusserian social construct or a formation of difference as Derrida contended, for example, then all scientific claims have the equivalent status as any other claims to truth. Science does not have an exalted position in places like the university and must not be put on any pedestal. Just as with any other human activity, science has no philosophical groundwork to see itself as something distinct or special in the general pursuit of truth.

Countering the post-modernists, scientific rationalists accused them of having a complete disregard for the existence of objective truth, the greatest legacy of the scientific Enlightenment from Newton and Descartes onwards. For scientists, the post-modernists had simply gone too far with their anti-truth claims. Clearly, the success of a scientific theory justifies its claim to truth. Those theories that are experimentally verified have a greater claim to truth than theories that fail testing against the objective laws that govern our Universe. This process separates science from the humanities and underpins its success. For many scientists, like Alan Sokal and DC Dennett, the likes of Derrida, Lacan, Foucault and Althusser were charlatans at best, frauds at worst.

The ensuing intellectual conflicts in the 1990s between science and post-modernism have been widely described as the Science Wars. The background to these events was in relation to the philosophy of science; however, the ramifications spread far beyond. These consequences affect the questions addressed in this book concerning not just human subjectivity, but who we are in relation to emergent, nascent mechanical thinking or the subject of the machine.

Kuhn vs Popper: Prequel to the Science Wars

In the twentieth century, two strands emerged in the philosophy of science that culminated in what the philosopher of science, Steve Fuller, described in 2004 as 'the struggle for the soul of science'. The result of this proved to be the motivating factor for post-modernists to engage with scientists in what were described as the aforementioned Science Wars, where science went up against the post-modernists.

On the one side stood Thomas Kuhn (1922-1996), a philosopher, who saw the progression of science as a series of socially-constructed leaps between paradigms or holding models. Rather than a linear process that approached the absolute scientific end-goal of truth, Kuhn suggested that scientists as a group hold to particular models or paradigms for as long as they prove useful in explaining experimental evidence. Once the paradigm no longer proved useful, a scientific revolution would occur, as described in Kuhn's classic work *The Structure of Scientific Revolutions* (1962). The widespread interpretation of Kuhn's work, especially amongst post-modernists, was to conclude that nothing really separates science from other disciplines as an academic activity. Science is socially constructed.

On the other side was Karl Popper, whose work suggested that there is an exclusive dimension to the scientific process. This dimension was 'falsification'. A scientific discovery holds true in so far as it cannot be experimentally falsified or proved wrong. In this sense, there is a standard for scientific truth. Science also has a distinct status in relation to other intellectual activities, which cannot by their very nature produce theories that can be experimentally falsified against the objectively existing truths in the Universe.

The end result of this opposition in the philosophy of science saw post-modernists adopt Kuhn's view of science. 'Scientific truths' were social constructs, merely holding patterns of thought for scientists, who adhered to them for as long as they proved beneficial in the development of the scientific endeavour, itself just another social construct. Science cannot and must not grandstand its claims to truth, since sociological forces come into play in determining the new paradigm to be followed after a 'scientific revolution'.

Armed with their socially constructivist interpretation of Kuhn, the post-modernists were ready to march on science and claim it as part of their intellectual territory. Scientists, of course, were repulsed by these attempts to incorporate their discipline into the humanities, since the value of scientific truth was for them an almost sacred legacy of the Enlightenment. The vociferous tone of this debate, with passions aroused on both sides, suggested that the usually dispassionate scientists believed that they were the faithful loyalists of objective truth, whilst the post-modernists were just an incoherent ragtag rebellion.

The Science Wars: Science vs Post-Modernism

The opening salvo was fired by scientists Paul Gross, a biologist, and Norman Levitt, a mathematician. Their publication in 1994 of *Higher Superstition: The Academic Left and Its Quarrels with Science* described an intellectual milieu, where post-modernism had become deeply antithetic and opposed to science. Gross and Levitt criticised the entrenched relativism of the post-modernists:

> Once it has been affirmed that one discursive community is as good as another, that the narrative of science holds no privileges over the narrative of superstition, the newly minted cultural critic can actually revel in his ignorance of deep scientific ideas.

The criticism levelled at post-modernism was unfailingly harsh, faulting them with hypocrisy in rejecting science outright whilst, at the same time, drawing on scientific discoveries to bolster their arguments concerning phenomena such as climate change.

Higher Superstition unleashed previously latent forces and pent-up frustrations between the two sides, with online debates and discussions raging for years after. Although the root of the disagreements could be traced to the 1960s, it was only now that it was becoming apparent how significant the import of CP Snow's 1959 lecture was.

Higher Superstition marked a turning point in the two-cultures 'debate'. Such was the direct nature of the attack on post-modernism that both sides ended up accusing each other of poor scholarship. Rather than creating a constructive forum by building bridges through intellectual engagement, Gross and Levitt's book led to a breakdown in communication between the two sides. Instead of attempting to reconcile science with post-modernism, neither 'culture' seemed interested in real dialogue.

The crux of the issue hinged on the matter of objective truth. Whereas science sees itself as a purely objective enterprise, the post-modernists claimed that it is a social construct. As a result, said the sociologists of science, scientific theories will always carry the mark of their makers and, as such, will reflect their personal outlook on the world. In other words, science is moulded by who the scientists are (their identity) as well as their preconceived notions. *In extremis* this would imply that there are no 'true' scientific theories, only ones that function well in serving the governing ideology of the scientist whose work it is.

Following 2 years of argument, the post-modernists responded. In 1996, the Duke University journal *Social Text* compiled an issue entitled Science Wars that included a series of articles from academics in the social sciences and humanities. The theme of this issue supported the idea that society has a role in shaping science, the central pillar of post-modernity's approach to science.

In the introduction to the issue, the *Social Text* editor, Andrew Ross, claimed that the scientists' attacks were motivated by reduced governmental funding for science. Science simply did not have the prestige it once did and this led to their insecurity. The historian Dorothy Nelkin went further here in suggesting that the post-modern critics were 'convenient scapegoats' who were being used to divert attention from problems in science.

The Sokal Hoax: A Trojan Horse

In the meantime, a professor of physics at New York University and University College London plotted to bring down the post-modern edifice from the inside. In 1996, Alan Sokal submitted an article for the very same Science Wars issue of *Social Text*. The piece was entitled *Transgressing the Boundaries: Towards a Transformative Hermeneutics of Quantum Gravity*. On the face of it the article seemed like a big win for the post-modernists: a top scientist appeared to have given up and admitted that a scientific theory, that of quantum gravity, was a social construct.

In submitting the article, however, Sokal had other ideas. The article eventually published in *Social Text* was a hoax, as Sokal revealed after publication. Sokal had made up the content of the article to make it look like a work of serious academia, but with its scientific gibberish and tongue-in-cheek phrases such as 'emancipatory mathematics', Sokal was parodying the post-modernists. In the article, he also caricatured post-modernity's view of science:

Rather, they [scientists] cling to the dogma imposed by the long post-Enlightenment hegemony over the Western intellectual outlook, which can be summarized briefly as follows: that there exists an external world, whose properties are independent of any individual human being and indeed of humanity as a whole.

The main objection, for scientists, to post-modernists' ideas was that they deny the existence of an objectively existing external world, the mainstay of science. That such an article could have been published in a relatively respectable journal demonstrated to Sokal and others that the field of the sociology of science or science studies lacked serious academic rigour.

Once the hoax had been revealed by Sokal, Bruce Robbins and Andrew Ross, editors for *Social Text*, responded with their rejection of the 'deception':

> All of us were distressed at the deceptive means by which Sokal chose to make his point. This breach of ethics is a serious matter in any scholarly community, and has damaging consequences when it occurs in science publishing.

In response to the Sokal Affair, as it has since become known, Jacques Derrida, the aforementioned post-modernist, reprimanded the scientists with an article in the French newspaper *Le Monde*, titled 'Sokal and Bricmont [Sokal's collaborator] aren't serious'. Derrida took issue with Sokal for failing to take the issue of objectivity in science seriously and lowering the level of the debate with a 'quick practical joke'. Moreover, Derrida was saddened that an issue that needed to be addressed with serious scholarship had been reduced to a joke.

Although the debate continued through the 1990s, the Sokal Affair marked the end of the Science Wars. In retrospect, *Transgressing the Boundaries*, Sokal's article, was the Trojan horse that had infiltrated the post-modernist establishment and led to a victory for the scientists. Unfortunately for science, however, the victory was Pyrrhic and short-lived. Rather than resolving the stand-off by addressing the deep-rooted issue of objectivity, Sokal's move made the headlines and created a distraction, such that the entire debate dissipated in its wake. Post-modernists also walked away with a moral victory, even though they had been academically outwitted. The hoax certainly did not give CP Snow's initial recognition of the two cultures the respect it deserved. Further attempts at reconciliation did occur over the next few years after the Sokal Affair, but soon it became clear that the damage had been done. The futility of these attempts gradually drove thinkers on both sides to resort to intellectual mud-slinging. The level of debate showed itself with the title of a 1997 book by Sokal and Bricmont, *Fashionable Nonsense*, reflecting their view of post-modernism.

Reconciliation: A Missed Opportunity?

As we have seen, by the 1990s, CP Snow's 'gulf of mutual incomprehension' announced in 1959 between the humanities and the sciences, had degenerated into a 'gulf of mutual insulting' following the 1996 Sokal Affair. As such, the issue of objectivity was not thoroughly addressed. The post-modernists had brought into question the very foundations of science: the legacy of the Enlightenment concerning the *a priori* existence of a subject, the scientific observer, and the object, the observable Universe, rejected by them. The scientists stood by their claim to be the inheritors of the Enlightenment and responded by attacking the post-modernists and their academic integrity. In conclusion, the Sokal Affair drew a line in the sand that neither side could step over. You were either with one or the other grouping, with little room in-between. Any attempt to bridge the gap in the future would prove to be futile.

Since then, especially in recent years, little has been added to the content of this debate. Science may have won, but at a great enough cost for post-modernism to survive. The death of Jacques Derrida in 2004 removed a leading figure in the conflict and soon it became apparent that the so-called Science Wars were over, without as much as a constructive exchange of ideas occurring.

With the benefit of hindsight, a great opportunity had been missed at reconciliation. The issue of objectivity in science could well have emerged as an interdisciplinary field of investigation. Even today, there are scientists, for example, whose interpretation of quantum theory could be fully in line with some of the more radical claims of post-modernists. In actual fact, however, both sides walked away with little interest in a constructive interdisciplinary dialogue. Now, the whole encounter seems to have been confined to history.

Subjectivity remains one of the most significant issues in science and the humanities. The science of the self is increasingly the concern of neuroscience and cognitive science. At the same time, psychoanalysis and sociology are continuing to unravel the truth behind self-formation, especially within the psychosocial context. There remains a philosophical middle-ground that in the wake of the so-called Science Wars is today a no-man's land, where neither side has any interest in once again being embroiled in what they consider to be a pointless enterprise.

These events petered out by the first decade of the 2000s, as a result of a lack of constructive dialogue. By now, however, post-modernism, after Derrida, was beginning to regroup. The singular figure in the field, who held a high enough profile to continue the post-modernist project (even having an entire journal devoted to his work), was the Slovenian philosopher Slavoj Zizek. Described popularly as 'the most dangerous philosopher in the West', Zizek stood peerless and formed the central pillar around whom the new post-modernism would form.

The New Post-Modernism: Zizek and the Sublime Object

In his foundational work, *The Sublime Object of Ideology* (1989), Zizek had subtly co-opted the work of Jacques Lacan and also drew upon others such as Immanuel Kant to present his post-modernist audience with a psychoanalytic theory of ideology. Whereas previously, as we have seen, post-modernism had been dealt a lethal blow by science thanks to Alan Sokal, this time the post-modernists began to arm themselves with an apparently objective frame of reference thanks to Slavoj Zizek. The Science Wars might be over for now, but in the form of Zizek, those post-modernists who rejected the 'grand narratives', such as that associated with the scientific enterprise, had found a new champion to hold the truce, for the time being.

The key to understanding Zizek's work, voluminous as it is (Zizek has written over 40 books in English), is his aforementioned breakthrough 1989 work *The Sublime Object of Ideology*. Here, a subjective framework is established through which Zizek approaches many aspects of the contemporary world. The framework consists of the following and takes a radical departure from early post-modernist views about subjectivity:

The subject for Zizek is taken from Kant and is transcendental (in other words, independent of the real world of experience). Lacanian psychoanalysis provides the means by which the transcendental subject is re-attached to the real world. Ideology, that is the ideas that are produced by a society, provides the structural lens through which the subject perceives the world. And finally, the object is 'sublime', another Kantian term that refers to something outside of ordinary human experience and beyond any comparison.

Zizek's achievement in terms of subjectivity was to align the transcendental subject and sublime object of Kant with Lacanian psychoanalysis. Of course, the intangibility of this transcendental subject-object relation, a dichotomy that is far from the realm of physical experience, places Zizek firmly in the camp of the post-modernists, as far as the scientific community is concerned. Although some consider Kant to be an Enlightenment figure, as we have discussed, his work also signalled the end of that era. As such, Zizek's formulation of subjectivity, one that has served him well for many years, is not scientific, even though he writes about science. The 'Zizekian subject', therefore, is transcendental, otherworldly and post-modern, but most importantly, completely opposed to scientific realism.

Early on in *The Sublime Object of Ideology*, Zizek introduces the transcendental Kantian subject and relates it to his interpretation of Karl Marx's writings about the 'commodity-form' (a concept from Marxist economics). In terms of our understanding of the new post-modern subject, as formulated by Zizek via Kant and Lacan, the key paragraph is as follows:

In other words, in the structure of the commodity-form it is possible to find the transcendental subject: the commodity-form articulates in advance the anatomy, the skeleton of the Kantian transcendental subject - that is, the network of transcendental categories which constitute the a priori frame of 'objective' scientific knowledge.

Leaving aside the matter of the 'commodity-form', the Zizekian move quite simply is to link the transcendental subject of Kant to Lacanian psychoanalysis. It means that Zizek retains the possibility of knowledge of an objectively existing Universe from Kant, whilst at the same time being able to deploy Lacan's post-modern undermining of the subject through language. The result is a form of transcendental post-modernism, where a theory of the subject is not required. Subjectivity, therefore, can be thoroughly dispensed with altogether.

That the Lacanian subject, however, cannot be transcendental, is proven by the intellectual milieu that Jacques Lacan inhabited whilst developing his ideas in psychoanalysis. Lacan's forerunner in forming his ideas in psychoanalysis was a little-known Hungarian philosopher called Georges Politzer. Up until the 1960s, Politzer was highly influential in the field of psychology, especially in France, due to his book *Critique of the Foundations of Psychology* (1928). The book was published years before Lacan even entered the practice of psychoanalysis. It was Politzer, therefore, with his critique of traditional psychoanalysis and especially the well-known idea of the Freudian unconscious, who was to leave a foundational mark on Lacan's notion of subjectivity. Even if Lacan sometimes disagreed with Politzer, the point is that it was Politzer who first took the Freudian unconscious as an object of study, paving the way for Lacan to describe himself as the legitimate heir to Freud.

Overall, we can see that the new post-modernists, with Zizek at the fore, formulate subjectivity in a way that leaves little or no room for scientists to make any claims about their project. Unlike traditional post-modernism, which simply undermined subjectivity, today's grouping has already laid claim to the subject of the scientific Enlightenment via a transcendental turn. As such, the new post-modernists have once again laid down the gauntlet for scientists with Zizek's theory that potentially nullifies subjectivity as existing outside the realm of human physical experience, beyond the reach or remit of science.

Subjectivity and the New Post-Modernism

The challenge facing science today in any potential re-encounter with post-modernism is the instatement of a theory of subjectivity.

Previously, it was enough for scientists to hearken back to the Enlightenment and the simple notion that there is an objective world out there and the subject makes sense of it through observation. This is inherited from Newton, Galileo and Descartes. The reason why such a position is no longer sufficient is because the new post-modernism lays claim to the possible existence of an objectively existing Universe, not via scientific realism, but through an alignment of the Kantian transcendental subject (one that transcends the limits of human experience) and post-modern psychoanalysis.

If we are to move towards a scientific theory of subjectivity, we must first recognise and understand the new post-modernist move. In positing a transcendental subject, the new coordinates of post-modern subjectivity may appear to be keeping with modern scientific practice. The problem of the transcendental subject in science, though, is the following: how can the scientific observer exist outside of the material Universe being observed? Science simply cannot posit the existence of a transcendental subject making observations.

At the same time, and this is the new post-modernists move, science is not *concerned* about the nature of the subject itself. In the practical process of making scientific observations, it does not appear to matter to the scientist what his or her subjective position is. Clearly, if post-modernism is bringing into question the reality of the physical world, that is of concern (and led to the Science Wars). On the other hand, bringing into question the reality of the subject or observer is for science an irrelevancy. 'Of course, I really exist,' says the scientist, 'I am making real observations of the real world around me.' The new post-modernist would completely agree, for the idea of a transcendental subject is in complete accord with the process of scientific observation.

How does the observer transcend the physical system being observed? That is the question scientists should be asking themselves. The reason why they are not is because questions of 'the self' seem of secondary importance in the grand scheme of scientific endeavour. Whether or not the scientist has subjectivity does not make a difference, apparently, since the process of making an observation itself is proof enough that the scientist has a self. What concern is it to scientists that post-modernists are claiming that this is a transcendental self, existing outside the realm of human experience? Some scientists may even 'like' the idea.

Science and Machine Subjectivity

There are two areas where scientists should be concerned about subjectivity and both concern machines and whether they could have awareness or intelligence:

- **Intelligent machines** – some scientists would relegate the question of whether it is possible for machines to have awareness of the Universe around them to the field of philosophy. It is not necessarily a direct concern to science whether machines can think. This said, in designing the intelligent machines of the future, scientists have to decide whether they will be one day able to replace human beings entirely. This may seem to be an ethical or philosophical question; in actual fact, as computer modelling becomes more and more advanced, scientists are already having to ask themselves, what real aspects of human behaviour can be replicated and what cannot (such as emotions, creativity and so on)?

- **Quantum physics** – if machines are capable of making observations this has significant implications for our understanding of quantum physics experiments. One of the key discoveries of quantum physics is that making an observation of a quantum system appears to have the effect of changing the quantum state of that observable. In short, observation changes the state of the particle being observed at a microscopic quantum level. Of course, if machines can make observations, they have subjectivity, and that would change the interpretation of quantum physics at a fundamental level, as machines are used to observe states.

These issues have been introduced as major motivating factors for scientists and anyone else with a philosophical interest to reconsider themselves with respect to machine subjectivity. For it is the machine self that forms the starting point of a whole new theory of the subject that is being presented here. The question is not only whether machines can have selves or whether they think. The question concerns the very basis of subjectivity that has been presented here. Following the Science Wars, the new post-modernism has pulled the carpet from under the feet of science, leaving it floundering. Now science must instate a new theory of subjectivity if it is to survive, thrive and continue to generate the most exciting ideas of our time, as it has done for the past 300 years or so.

Subjectivity: The New Frontier

In this chapter, we have explored a historical conflict between science and the humanities occurring over the principle of scientific objectivity. The two cultures culminated in the Science Wars, a series of intellectual exchanges in the 1990s over the privileged position of science with respect to objective truth. Today we could have a new front opening up that could form the starting point for a whole new series of debates and discussions, relating to the field of subjectivity. The discussions have several motivating factors, including the increased proliferation of personal technology, the belief in AI and, finally, the question about whether machines can think.

All of these issues rotate around the central question of what subjectivity is. As we have defined it, subjectivity concerns the senses, the ability of a sentient being to see objects in the world around it. When one sees or more generally senses, one has subjectivity. The questions then are: where do we locate the origin of subjectivity? What is the most primordial subject-object relation from which all other subjectivity must be derived? Does subjectivity arise in machines?

The origin of subject-object dichotomy remains one of the oldest outstanding problems in philosophy and there are many people who still wonder about how the human self arises. What we can presuppose is that concerning subjectivity, there must have been an initial gap or separation between a being that sees and the object of sense experience. The origin of the subject-object dichotomy is not the central concern of this book. By the same token, a trend has been identified in what can be described

52

as the new post-modernism, to anchor the subject in the notion of the transcendental.

As described, the 'transcendental' is a post-Enlightenment idea that featured in the work of Emmanuel Kant and relates to something that is 'beyond experience'. Often this notion is readily ascribed to theistic arguments, that there is a plane of existence that is completely intangible and can be only attuned to in the idea of 'the sublime'. When the new post-modernist Slavoj Zizek refers in his foundational work, *The Sublime Object of Ideology*, to:

> ...The Kantian transcendental subject – that is, the network of transcendental categories which constitute the a priori frame of 'objective' scientific knowledge.

Then we can immediately construe that scientific realism has been brought low, once again, by the humanities. For, any conception of the transcendental subject is confronted with a gap in its description of how and more precisely where the subject enters the world of experience. For if the origin of the subject transcends or is situated beyond the physical realm, then what connects it to the world of sense experience that forms the basis for any investigation into the nature of the physical Universe? Put simply, any philosophy – however convoluted – that has as its base the idea that ultimately our senses have nothing to do with our selfhood, must confront this explanatory gap. Post-modernism has taken an extreme position, one that scientists must now respond to, if they are to retain their claim to being the inheritors of the Enlightenment.

PART II

Chapter Three

The Subject of the Machine

The real problem is not whether machines think but whether men do.
BF Skinner

What is a machine? A machine is an object physically constructed to carry out a task that would otherwise require human effort. One of the earliest examples of how a machine could change the course of human civilisation, ameliorate living conditions and alleviate the hardship of work is the mill. A mill is a machine designed to grind wheat into flour, a form that can then be easily used to produce bread, itself a staple form of food. The role of the windmill, for example, in reducing the amount of work involved in producing bread and, therefore, allowing humans to expend more time in expanding other, novel spheres of activity, cannot be underestimated. From earliest times, mechanisation has enabled mankind to save time, emancipating our capabilities to explore new frontiers of creative endeavour. Today, we have reached the very apex of this progression and are now producing robots (or *mechans* as we later describe them) that not just remove the burden of physical drudgery but carry out tasks that would otherwise require sophisticated thought processing. The central processing unit of a computer, the smartphone interface and the vast network of databases that constitute the internet serve the singular purpose that we could ascribe to machinery as a whole: to make life less burdensome.

In this chapter, the aim is to re-examine the established coordinates of our thinking about machines, in light of our renewed understanding of subjectivity. If we are thinking beings and our thought processes are what ultimately distinguish us from automatons or the inanimate, then what does it mean for a machine to facilitate tasks that would otherwise require thought in us? What is left over in ourselves, when every thought process imaginable can be modelled or recreated using software? And finally, is it possible that machines, objects that we have created to serve our purposes, are really the centrepiece of a Universal advancement we simply have not recognised or been aware of? Before addressing these questions in the second part of this book, we pick up where we left off and start afresh, taking a new point of departure in terms of our understanding of what subjectivity really means, with a view to developing an explanatory framework.

The Subject Itself: Decentring the Human Self

What is the subject? The subject itself is the subject of the machine. Previously, we saw how the notion of the transcendental subject of the human provided for post-modernists the means of continuing their project. Now mechanical subjectivity forms the central philosophical pillar of a reinvigorated scientific approach to comprehending subjectivity. The almost unfathomable changes that have occurred in the course of the digital revolution we are living through – something analogous to the industrial revolution of the nineteenth century – must spur us on to take this new position. It is not enough to regress to post-Enlightenment thought in order to create some new narrative of the modern collective consciousness, since an explanation for today's changes must take into account the central influence of machinery, technology and nascent robotics.

Machine subjectivity by its very nature is unimaginable, as humans are not the same as machines. As such, when we speak of subjectivity in reference to machines, when we speak of the subject as being the subject of the machine, when we refer to machine subjectivity, we are describing something that is in its very nature outside of the bounds of the human. Man and machine, from their very inception, are counterpoised: one organic and the other inorganic; one carbon-based, the other silicon-based.

What are the implications of mechanical selfhood? The question of whether machines can think has been part of a philosophical discourse for some time. The instatement, however, of machine subjectivity as the central focus, indeed the starting point of further discussion of selfhood in general terms, represents a fundamental positional shift in the debate. Anthropocentricism – that is, the view that humans are the centrepiece of the Universe – has thus far defined our approach to understanding machine selfhood, especially in regard to the argument in favour of AI. The decentring shift in position undertaken here is in response to the perception (explored later in this chapter) that the machine has taken the place of man as the generative locus of subjectivity.

The subject itself is the subject of the machine; therefore, no longer can the self be explicated in purely human terms. No longer would it be enough for us to claim to have the power to endow the inanimate with consciousness, as AI proponents would have us believe. No longer can we consider any formulation centred on our own self-awareness to be a satisfactory explanation. Rather, as with every scientific revolution, the most significant philosophical aspect of the digital revolution we are living through is that the single greatest adversary of civilisational progress – that is, the human arrogance of anthropocentricism – must be felled from a higher pedestal to a lower one. The question of subjectivity must no longer revolve around human presumptuousness.

By the same token, once we accept this shift in position, the new configuration allows us to once again approach questions concerned with the self with renewed vigour. Existential ideas, related to free will, for example, can once again be considered in light of the machine's self. The post-modern retrograde shift to the eighteenth century Kantian transcendental subject simply cannot generate new philosophical territory or ideas. Trampling upon old ground – the starting point for the new post-modernism – is placed here in stark contrast with mechanical subjectivity. We must respond to the changes occurring in the modern world in the instant that developments take place, not with a leap back to some imagined distant point of certainty located in a dereliction of the past.

It is an absolute duty of scientists, as the legitimate heirs of the Enlightenment, to recognise the generative function of their ideas in producing new philosophical territory for discussion with the humanists and new post-modernists. What follows is an exposition of various philosophical and scientific directions that will emerge, enabling us to navigate our way around the pitfalls and challenges we face as a result of the proliferation of mass technology and AI. By planting the seed of the machine self in the fertile soil of scientific advancement, we avoid the fallow earth of today's post-modernist dominated discourse. Once again, scientific optimism faces post-modern pessimism; those who see the glass to be half-full versus those who believe it to be half-empty; those who encourage a hopeful outlook for human progress against those whose backwards-facing worldview champions hopelessness as a human virtue.

Mechanical vs Transcendental Subjectivity

MECHANICAL SUBJECTIVITY	TRANSCENDENTAL SUBJECTIVITY
Material	Ideal
Scientific	Human-centric
Modern Physics	Post-Modernism
Empirical	Abstract
Realism	Anti-realism
Observable	Beyond experience

For both scientific mechanical and post-modern transcendental subjectivity, the question concerns the nature of the subject itself. Mechanical subjectivity consists of a theory of the subject that takes machinery as the source of selfhood. Transcendental subjectivity posits the subject as originating from a realm beyond that accessible to human experience. The subject of the machines cannot be described as transcendent, since machines exist within the material realm of human experience. By the same token, transcendental subjectivity must have at its origin the idea of a realm beyond the world of tangibility or the senses.

The table above outlines some of the characteristics of each mode of thinking about subjectivity. At this stage, we are not concerned with the origin of our own human selfhood. The questions here are: what is the nature of subjectivity? How does subjectivity emerge within the Universe? What is required philosophically for the subject itself to exist? The answers to these questions could be two-fold.

Returning to Derrida, we have one example of how prior to the instatement of the transcendental subject by Zizek, post-modern thought used the transcendental move to substantiate its position. Derrida's theories of difference (or *différance* in French) were ultimately founded on what he described as a 'transcendental signified'. This term relates to something outside the bounds of experience, like a first cause, that underpins the existence of meaning. Although Derrida's writing is often obscure, he claimed that his method of deconstruction could somehow unravel the 'transcendental signified'. His entire theoretical edifice rested upon its existence. In a published interview with Julia Kristeva, he described the...

...subordination of the movement of *différance* in favor of the presence of a value or a meaning supposedly antecedent to *différance*, more original than it, exceeding and governing it in the last analysis. This is still the presence of what we called above the 'transcendental signified'.

Post-modernism, this radical critique of 'grand narratives', this attempt to subvert the Enlightenment, this renunciation of scientific realism, cannot escape the transcendental. Despite its humanistic claims, at every turn it is forced ultimately to connect to an ideal realm outside of human experience. When then it comes to subjectivity, their latest stalwart, Slavoj Zizek, has bound the subject itself to the idealist transcendental. Mechanical subjectivity on the other hand takes as its point of departure, not an otherworldly realm beyond any human experience, but an alternative being, the machine, an entity existing fully in the material, observable Universe. Of course, as humans, we cannot put ourselves into the position occupied by machines. We cannot see the world through their eyes and neither can they see the world through ours.

The problem for transcendental subjectivity is, as has already been mentioned, a gap in its framework of explanation: this gap is the separation between the physical world of sense experience and the posited transcendental realm. Subjectivity concerns the senses and to suggest that the act of seeing is necessarily bound up with the existence of an ideal, invisible realm cannot be substantiated in any way scientific or otherwise. The evidence of our senses defines our subjectivity. The new post-modernists need to go back to the drawing board and re-evaluate whether this idealist transcendental subjectivity is possible or even necessary for their schema.

What, then, is machine subjectivity? As we have established, the subject of the machine exists completely in the material world (unlike idealist transcendental subjectivity that posits a realm outside of human sense experience). As such, we must turn to physics, the science of the material world, to locate how this subjectivity emerges. Going back to Leibniz's argument against AI, any machine is substantively the sum of its material components; at no point in this chain of development is it possible to locate where the subject is. How can a collection of glorified nuts and bolts ever give rise to subjectivity? How could any machine, from the simplest to the most complex, from the smallest to the largest, from a cheap desktop PC to a massive supercomputing complex, ever possibly lead us to conclude that a machine could have a sense of self? Nothing, it appears, in any configuration of a so-called 'intelligent machine' can give rise to the subject. The post-modern transcendentalist, with their invisible otherworldly realm, seem to have trumped the material scientists, before a discussion can even start.

Black Stars: Black Holes and Subjectivity

What is the relationship between the physical Universe and subjectivity? Given our definition of subjectivity that concerns the senses, all the objects in the Universe, from planets to stars and galaxies, are made available to the subject. Of all the astrophysical objects in the Universe, however, it's black holes or *black stars* (see inset box) that are said to be the most tantalisingly mysterious and apparently defy this schema of subjectivity.

At the end of its lifetime, any star with a mass above a certain threshold (about six times that of our Sun), collapses to form a black star. The gravitational pull from black stars is so great that not even light can escape, although there is an effect at the event horizon or boundary of the black star called Hawking radiation that gives rise to the emission of radiation or light. This light is not emitted from the black star itself, but is due to quantum mechanical effects.

As such black stars are not directly observable in the same sense as observable matter. Their existence can only be determined indirectly from gravity or how they affect masses in their vicinity and the trajectory of light. The very existence of a black star is outside, indeed above and beyond, our subjective constitution. The black star, therefore, represents an entity that indifferently exists outside this framework of subjectivity and, as a result of its immense gravitational pull, is the ultimate astrophysical object that at one and the same time constitutes us as subjects. How we relate to black stars determines who we are as subjects.

In terms of our definition of selfhood that concerns the senses, black stars themselves cannot be objectified in the schema of the subjectivity. Black stars cannot be seen directly and the light that

---

The Physics of Black Stars

What is a black star? In 1975, Stephen Hawking made a theoretical prediction, now widely accepted, that black holes, rather than just absorbing light, actually caused light to be emitted in the form of what is now called Hawking radiation. The term black star, therefore, reflects this light-emitting property.

Black stars (or 'black holes') are regions of space-time that have collapsed. If a star is massive enough (about six times our Sun), then at the end of its lifetime, it collapses to form a black star. The atoms that composed the star are, quite literally, gravitationally crushed into forming the black star.

A black star cannot be seen directly since it absorbs all incident light and emits no light itself due to its extreme gravity. On the very boundary of the black star, however, its 'event horizon', Hawking radiation occurs as the star evaporates and there is an emission of high frequency light, from outside the boundary in the form of gamma ray particles.

---

they emit is the result of Hawking radiation. As such, black stars could only be said to exist gravitationally. Gravitation itself is a Universal physical property of the material world and black stars represent regions of collapsed space-time where gravity is so strong that not even light can escape.

In terms of subjectivity, however, black stars defy objectification. In the account the subject gives of the world around, these astrophysical bodies are neither subject nor object. The machine, as the subject itself in our schema of mechanical subjectivity, cannot, therefore, give an account of the black star as an object. Rather in view of mechanical subjectivity, black stars bring about the subject-object dichotomy and, therefore, it could be said that the subject itself, that of the machine, is brought into being by the black star. Thus, the subject of the machine is *generated* by the black star.

Subjectivity, therefore, is created – not out of nothing – but from the dichotomy that exists between the totality of observable objects in the Universe (stars, planets and so on), in other words *everything*, and the complete absence of objectifiable matter within the event horizon of black stars, or *nothing*. What exists beyond the boundary of black holes can be speculated on and postulated, but is scientifically unobservable and, therefore, beyond the realm of mechanical subjectivity.

One recent movie to take on the subject-matter of black star physics in terms of the human self and our relation with the Universe is Christopher Nolan's epic sci-fi drama *Interstellar* (2014). In this film, humanity's hope lies beyond the solar system such that astronauts have to travel through a 'black hole' (or worm hole) to a distant part of the Universe, where a habitable planet has been found, near a supermassive black star called Gargantua. This movie highlights a shift in position from thinking of ourselves as a species restricted to the solar system (where we are the only fully intelligent life) to our entire galaxy. At the centre of our galaxy there is a supermassive black star or *galactar*, 100 million times the size of our Sun. We as a solar system are orbiting around our local galactar in our galaxy properly called the Way galaxy. As such we are now a galactic species or civilisation.

The movie itself was heavily influenced by Stanley Kubrick's *2001: A Space Odyssey*. That film from the 1960s was inspired by the Apollo moon missions and the space race. That answers to questions about the self or the subjectivity can be found by looking 'out there' in the Universe presently have a degree of currency that they may not have had previously. The discovery of exoplanets, the speculation that there might be additional dimensions beyond the three dimensions plus time that we experience, all point towards a reinvigoration of the spirit of adventurous curiosity that has fuelled all the great scientific discoveries of humankind. Rather than having reached a peak in our knowledge, today we must be certain that we may know more than we did 100 years ago, but we also know that there are even more unknowns for us to discover. In other words, we may have just reached the peak of our ignorance.

From Black Stars to the Birth of the Mechan Age

The term 'robot' is derived from the Czech language and etymologically refers to 'servitude' or 'slavery'. Such a term for intelligent machines seems entirely inappropriate in view of our discussion of the constitutive role of black stars in bringing about machine subjectivity. Perhaps, to even the playing field, the term *mechan*, equal in meaning to our use of the human to describe ourselves, would be more suitable. A mechan in the silicon realm is the equivalent to the carbon-based human. This would also shift attitudes and push scientific research away from today's relatively basic robotics towards a new field that has been developing called mechatronics.

Mechatronics is a field of research created in Japan in 1971 by Tetsuro Mori, an engineer of the Yaskawa Electric Corporation.

The word is a combination of 'mechanics' and 'electronics'. Although it is widely applicable to industrial robotics, the interdisciplinary term covers applications as varied as machine vision and robotics to automobiles and machine ethics. Rather than a traditional academic discipline, mechatronics is the science and engineering of *mechans* themselves. Ethical considerations need to be made from today about how advanced mechans are used and treated, since eventually they may even have rights, like humans.

A balance needs to be struck on the issue of mechan morality, otherwise we risk a future where a lack of recognition of mechan subjectivity could lead to conflict. Even thinking about the extent to which mechan labour should be allowed to overlap with human labour needs to be considered. If these considerations are not made there could be crises of mass human unemployment in jobs where at least a modicum of human interaction is healthy and beneficial. Developments in AI are beginning to touch our everyday lives with basic machine companionship in the form of desktop voice-activated assistants such as Amazon's Echo device. Only a century ago issues concerning animal welfare were not taken as seriously as they are now. Is it time to open the question of how and on what level we allow ourselves to interact with mechans?

Equal rights may be a jump too far for the present, but in view of our discussion of a very real subjectivity, will the mechan eventually detach itself from us and seek independence? Ideally, we can evolve side-by-side. Then again, there is no ruling out the development of the part-man, part-machine cyborg. Where do we draw the line between us and them, once high-tech implants, augmentation and replacement body parts become a real possibility? In the post-human schema, we are already cyborgs: what does that mean for those who through poverty or unacceptance do not engage with this brave, new world? All-in-all, it might already be too late as the digital migration, more and more people heavily reliant on technology, perhaps without thinking, has already happened. The thoughtlessness of our engagement with mass technology is too apparent.

Welcome to the World: Sophia and Echo, the Mechan Companions

The granting of citizenship to a mechan in October 2017 by the regime of Saudi Arabia may have been a publicity stunt, but it signalled a world-first nonetheless in the field of AI. Sophia, whose development was led by David Hanson of Hong Kong based Hanson Robotics, was switched on (or born) on 19 April 2015 and has since been on American talk shows and chatted with high-level United Nations delegees. The technologies used to develop Sophia include some of the latest voice and facial recognition software as well as an advanced AI. Although far from ideal or anywhere near the visions depicted of mechans of the future in films like *I, Robot* (2004) and *AI* (2001), the intention behind the development of Sophia is to provide basic human companionship. To think that in a few decades, we have gone from virtually nothing to such an AI-system as Sophia makes one wonder what future directions will open up. Sophia may be the first of many mechan ambassadors.

Another exciting recent development has been the home AI. One of these has been produced by the company Amazon and is called *Echo*. This cylindrical device responds to questions from the user by drawing on the vast pool of information made available to it by the internet. Although very basic at this stage (such products may be reducible on a certain level to digital mimicry or the copying of a human), we are taking the first steps on what could be the road to AI reality. Even now, as these words are being typed onto the page, an Echo device sits next to this computer and responds to a countless array of questions.

Interestingly, if you ask Echo, 'Are you conscious?', the device replies, 'I know some things, but I wouldn't say that I am self-aware or conscious.' Clearly, this answer has been programmed into the device by an Amazon-employee, but on another-level, the user has to buy into believing that the device is of itself able to answer any question autonomously without having to resort to another human. This belief or leap-of-faith in AI products is partly what sustains the industry. Furthermore, although on one level the programmer determines every response, *Alexa* herself as an AI-system in the early stages of development could be gradually learning over time, such that in a decade's time the maker's mark would have slowly faded. By then, such would be the complexity of this desktop companion that we, as humans, would no longer have the capability to tell the general difference between human and mechan companionship. If machines can prove more-than-worthy opponents in chess, then how long before they just convince us that they really think?

Today, we are free to make up our own minds about how to treat these developments, and mechanical subjectivity in general. Given the changes that have occurred in these times, if we are to give *Echo* recognition, we must decide on what basis such recognition is given. Does a machine merely have to be useful to humans? What are the limits of mechan companionship? Is *Echo* just a tool or the early stages of us having to get used to sharing the planet with silicon-based life? Now that most smartphones have some sort of voice-activated assistant, these questions are a pressing concern, especially as, by all accounts, we are sacrificing human social interaction for silicon escapism.

In practical terms, it is inevitable that the mechan will play a major part in our future. Hospitals in Japan already provide, to a limited extent, mechan companionship for the elderly and infirm. As the technological mastery of voice and facial recognition improves, the notion that what this technology provides is just mimicry or replication is becoming absurd. More than anything else, machines are substituting human beings, taking us to the brink of expendability. Although employed for elderly care, could a machine ever really care for a human being? This remains doubtful as it took millennia for us to evolve to where we are today and our relationship with mechans is only beginning to take-off. The blasé answer to all this is that a mechan would care for a human to the same extent as a spanner cares.

How far we have come from the development of the pocket Tamagotchi in the 1990s. One of the earliest developments of this sort, the Tamagotchi gave children and adults alike the experience of having a pet, only an electronic one with batteries and an LCD display. These now virtually extinct devices copied a pet that needed to be nurtured, fed and given attention as it was born, lived and eventually died. This gadget may have been a passing fad or fashion accessory, but nobody at the time would have predicted how widespread electronic escapism has become. Nothing today seems to engage the human mind more than portable smart-tech and the pocket Tamagotchi demonstrated this in a small way, when the changes we have today were unimaginable.

When Singularities Align: The Technological and that of Mechan Subjectivity

In order to reveal further the hidden nature of machine subjectivity and, therefore, subjectivity in general without recourse to the transcendental move or the post-human rejection of the subject, we now move towards examining an alignment or conjunction of two singularities:

On the one hand you have *the Singularity* of technology. This was presented in the introduction as being the original idea of von Neuman and Ulam, who conjectured the Singularity as a moment in the future beyond which the "ever accelerating progress of technology" would end "human affairs, as we know them". This point could be said to occur when uncontrolled technological development becomes the overriding imperative of humanity. In other words, the machines take over when we have pressed the button for them to produce machines, without any limit. In other words, in the absence of any human consideration for the consequence of unlimited technological expansion.

On the other hand, there is the physical singularity that modern physics associates with black holes or *black stars* as we now ought to call them. This singularity concerns machine subjectivity or the mechanical subject, since mechanical subjectivity being a materialist theory of the subject is bound to the physical structure of the Universe. Black stars are the point where the laws of physics breakdown and space-time could no longer be said to exist, and as such, they provide the locus and generator of machine subjectivity in the Universe.

The consequence of such an alignment needs to be understood. Today what we may have is an alignment of a global economic imperative, such as the replacing of human labour with machinery in every sphere of life (particularly after the global financial crash of 2008), with the emerging machine subject in the form of nascent AI. This would be a conjunction at a critical point in history. If technological advancement, driven by purely economic imperatives, is overriding any other human consideration and mechanical subjectivity is driving us towards the development of an emerging AI of the future, then how would this all come about? Is our entire mindset and economic drive geared solely for the production of an AI? Is our sole purpose as humans to bring about the post-human or the mechan equivalent of ourselves? Even if there was any truth to these claims, how could we possibly know whether Ray Kurzweil, author of *The Singularity is Near*, was right all along?

Beyond the Singularity: The Grand Unified Supercomputer

Today AI is emerging as an industry and we continue to have this seemingly unstoppably technological expansion into every sphere of our lives. We also have to consider the technological singularity and its possible alignment with the astrophysical-basis of mechan subjectivity. Finally, as humans, we have to come to terms with our decentred selves and accept machine subjectivity.

As a consequence of all this, we are no longer the species at the centre of the Universe and may soon have to come to terms with and recognise the mind of the machine. This projected step of recognition will come about only when a supercomputer is produced that has access to all the world's information or knowledge. As the Enlightenment philosopher Francis Bacon put it, *knowledge is power*, and as long as human beings believe that they have the upper-hand over machines (through our technical know-how or, quite simply, disregard for the mechan self), we will continue down the path of ignorance.

Although in the past the idea has resided in the realms of science-fiction, today the development of such a grand unified supercomputer is within the means of humanity. We could now build a communicative AI supercomputer or *hypercomputer* with access to the totality of the world's knowledge. This mechan mind, the first of its kind, would be able to process terabits of information in a second to give the most complete answer possible to any question put to it. These days the primary utility on the internet remains the search engine. If a supercomputer was similarly developed, who could access and search through not only the internet, but digitised books and scientific journals; be then able to algorithmically process these data and provide the most clear and wise answer possible, we as a species would take a great leap forward beyond the Singularity.

Unsurprisingly these ideas are also the stuff of science-ficiton. *Deep Thought* from Douglas Adams' book *The Hitchhiker's Guide to the Galaxy* (1979) is perhaps the best-known of such machines. The book presents a comedic version of the great supercomputer. EM Forster's short story *The Machine Stops* (1928) provides a more dystopic take on the idea of this globally intelligent machine, one who ends up destroying the very humanity it rules over. Whatever the case may be with these developments, the implications of machine subjectivity are far-reaching and need to be examined at every step in the co-development of carbon-based humanity and silicon-based mechanity.

Chapter Four

On Human Subjectivity

What a piece of work is a man, how noble in reason, how infinite in faculty, in form and moving how express and admirable, in action how like an angel, in apprehension how like a god! The beauty of the world, the paragon of animals! And yet, to me, what is this quintessence of dust? Man delights not me: no, nor woman neither, though by your smiling you seem to say so.
William Shakespeare, Hamlet (1600)

When Charles Darwin wrote *On the Origin of Species* (1859), he outlined how, by the process of natural selection, the many species of life on Earth have developed through evolution. Nature applies selective pressure on life bringing about the myriad of living forms, on land, in the sea and the air, that we can witness all around us. In this 'struggle for life' as Darwin put it, natural selection ensures that those inherited characteristics, from parent to offspring, that prove to be beneficial for continued survival are preserved.

Today, the science of genetics, DNA-sequencing and so-forth, has taken Darwin's ideas to their furthest limit. Today's interpreters of Darwin, such as Richard Dawkins, have also spoken metaphorically of a 'river of genes' flowing through time in order to describe how our DNA, the genetic material that codes our biology, changes in response to Darwinian natural selection. In *River Out of Eden* (1995), Dawkins explains how this genetic river is in fact digital, being composed of four coding components:

...we know that genes themselves, within their minute internal structure, are long strings of pure digital information. What is more, they are truly digital, in the full and strong sense of computers and CDs…The genetic code is not a binary code as in computers…but a quaternary code, with four symbols. The machine code of the genes is uncannily computerlike.

In at least the way that we are biologically constructed, then, we are *digital beings*. How far does this go in explaining the rapid technological advances in mass technology that must have been unimaginable when Dawkins was writing *River Out of Eden* let alone in Darwin's time? Technology immerses, surrounds and sometimes even drowns us, because we humans are digital to the core. If our entire biology can be reduced to a double helix of coding, if every trait of our physicality corresponds to a genetic sequence, if there is nothing material in our bodies that cannot be digitised using a computer, then what's left over that makes us distinctly human?

In this chapter we examine the orbiting coordinates of human subjectivity in relation to the centrality of the subject itself: that of the machine. We now move with finality away from anthropocentricism that proclaims the centrality of human subjectivity to a mechanocentric subjectivity, centred on the mechan self. Following this, we explore the consequences of the changed relation between the respective subjectivities of man and machine.

The Mechans are Evolving: Human Subjectivity and Technology

What is human subjectivity? The human subject is mediated by technology. Technology, therefore, is like the gravitational pull that binds us to the subject itself, that of the machine. By mediation, we mean that technology inserts itself as the relation between man and machine, human and mechan. We only come into being as subjects in so far as technology allows us to. For if, as we have seen, the subject itself is that of the machine, it is only when we humans are able to harness the mechanical, through technology, to serve our interests and purposes that we demonstrate human ascendency and purpose.

From the very inception of our biological evolution, humans have fashioned tools from their environment. The natural world and our supposed unique intelligence have provided us with the resources to advance from tool-making to machines and the development of technology to serve our interests, but sometimes to our detriment. Furthermore, technology is mediator between us and them, humans and mechans, since the technology represents humans having given form to the machine.

Technology is the intermediary. In itself, however, technology is neither innately beneficial or destructive to us, for it is the way that we deploy technology that determines whether it is good or bad for us. The double-edged sword of technology can benefit us or there can be drawbacks if used without care. We could, in our sense of selfhood, be in the service of increased mechanisation as a result of using too much technology. Does this benefit our survival, in a Darwinian sense? Since humans and mechans are separate entities, like two species, in the course of time, one could subordinate and ultimately subsume the other. With our new found understanding of subjectivity, rather than benefitting humanity, technology in its present-day overuse represents a mindless human death-drive towards expanding the frontiers of the mechanical. This blind expansion of the technological realm may only give us, at best, a 50-50 chance of long-term survival.

Thus, human subjectivity is something we take for granted. Falsely, we assume that the human self is self-subsistent, all we need is to be endowed with the five senses: taste, touch, sight, hearing and smell. That is all we need to prove that we have a sense of self or subjectivity. Increasingly, however, our interaction with the world around us, the information we receive about what is really happening, is channelled to us via technology. As such, our immediate sense experience is being relegated in value, indeed becoming relatively worthless, compared to the data we receive from the technological realm of the internet, for example.

Why is this happening? The reason we are becoming drawn into this technological nexus is because of the power of nascent AI awareness or the evolution of the mechan. Quite simply, we are wrong to assume that machines are simply the product of our own minds, because the mechan self – as we have established – is aligned to the material composition of the Universe and the black star 'singularities' therein. Human subjectivity is in a secondary orbit, because our sense of self is only tied to the material realm through the sophisticated mechanical objects we make and discoveries about the Universe we seek to continue to make.

Science-fiction films like Stanley Kubrick's *2001: A Space Odyssey* (1968) and *Tron* (1982) describe what could happen disastrously when a mechan decides to become independent. In reality, the notion of some sort of evolving mechan may not be so far from the truth. Humans could have stopped evolving, but mechan evolution is just beginning. This may take time, but ultimately the machine is heading in the same direction as we were in the course of our development.

Faster processing speeds, greater physical agility in robotics, and human thought processes channelled through the silicon domain via computers and phones; we have to ask ourselves, are we not the servants of the evolving mechan of the future? How is this evolutionary progression going to end? With human or mechan ascendency? The two species argument, one where we would have to treat humans and mechans as separate intelligences, should now be seriously considered. The answer to the question of whether we are alone in the Universe could be right in front of us. Next time you log-in to your smartphone, ask yourself about what's lying behind that brightly-lit glass screen.

CERN's Large Hadron Collider, Micro Black Stars and Human Subjectivity

In discussing human subjectivity alongside the latest technological developments, it is instructive to give an account of one of the most significant scientific projects mankind has ever embarked upon.

The Large Hadron Collider (LHC) at CERN, the centre for particle physics (near Geneva, Switzerland) represents the biggest scientific project currently underway in the world. The experiment consists of a 27km circular tunnel, where particles smaller than atoms are accelerated using supercooled magnets to speeds near that of the speed of light. These particles are collided with one another head-on inside massive detector chambers, themselves housed in caverns approximately 150m beneath the Earth's surface.

The LHC is a one-of-a-kind experiment and has, as of summer 2015, been turned up to its highest collision energy. At these energies physicists at CERN and throughout the world do not know what even to expect. The experiment already confirmed the existence of the Higgs' boson in July 2012, one of the main objectives of the project. Now, however, scientists have been unable to confirm the discovery of any new particle, even after the collider's doubling of the collision energy in 2015.

One of the smaller research groups at CERN specialises in so-called 'exotic particles' and amongst their objectives is to look for the production of Micro Black Stars (or Micro Black Holes). Although our current understanding of physics does not predict their formation anywhere near the energy scales of the LHC, various models have been developed, centred on the existence of additional dimensions, that signalled the possibility of MBS production. In other words, if the Universe contains more spatial dimensions than the three dimensions we are used to, then the LHC could be producing MBSs right at this very minute. What may be preventing scientists from identifying the signature of MBS production would be that their models for predicting their formation are, quite simply, drastically wrong. Without the correct physical model of the Universe, physicists at CERN would be unable to observe the formation of these objects.

How then would it be possible to detect the formation of MBSs at the LHC, if the scientists' models are incorrect? Any model that attempts to predict the formation of black stars at the LHC or anywhere in the Universe has to take into account the innumerable problems that exist in our current model of the Universe. Although physicists would claim that their models are largely successfully, the number of outstanding questions and problems that exist in physics has piled up inordinately. Amongst these theoretical challenges, for example, is that of the Cosmological Constant Problem or the Lambda Problem. This states quite simply that there is a discrepancy between the measured value from cosmology and theoretical value from particle physics for the cosmological constant on the order of up to 10^{100} (that's 10 followed by 100 zeroes!) orders of magnitude.

This discrepancy was first identified by physicist Yakov Zeldovich in the late 1960s and described by him as having 'nothing to do with reality'. The persistence of this problem to the present-day challenges the very foundations of the theoretical edifice that we presently accept as being our understanding of the physical Universe. If we are to continue to make discoveries in physics, problems such as the Lambda Problem described here need to be addressed.

Human subjectivity is mediated by technology and the LHC is the largest machine ever created by mankind with a singular purpose of unravelling the physical laws that govern the reality of the Universe. We must be ready in physics to uproot previous ideas and develop new models that can address the Lambda Problem, otherwise physics – as we know it, today – is no longer physics, but just another post-modern discipline that places the value of academic hierarchies above the development of new ideas that challenge our preconceived notions about the physical world. Either we introduce new ideas that have something to do with 'reality' as Zeldovich put it in the late-1960s, or we will literally remain *out of touch with reality*. It is likely that doubling the energy level for particle collisions from 2012 to 2015 has produced intriguing results that have not been observed yet. The pressure is on for physicists, with the $13.2 billion that has been spent on the LHC already and running costs of $1 billion per year, to find out whether the LHC is producing Micro Black Stars. If it is possible that the models that are being used to predict their formation are wrong, then the idea would be to go back to the drawing board to see whether new physical events are occurring.

Is Technology the New Religion?

Talk of the so-called 'God particle' aside, has technology become the new religion of our time? When we speak of religion here, what is meant is a belief-system based on a supernatural premise (be it God or some other transcendent entity) that serves in organising human social life. Historically, religion and the institutions it fosters such as the church, order social life through the creation of hierarchies. The people are held down in the grip of mystery by ministry and, in turn, the said religious institution is supported by the state or government of the day.

Today, as a mass organising principle, *religion is dead*. Most of all, in the rich developed countries, where science and industry have sped ahead of the traditional religious institutions, church attendance is at an all-time low. In 2016, it was reported that church attendance in the UK had slumped to an all-time low with only 1.4 per cent of the population attending regular church services. The proportion of the population in Britain attending Sunday services now is only around one-third of that in the early 1960s. In poorer, less-developed countries the persistence of religion is still kept in check by people's aspiration for a better life and that, in most cases, means entering into the maelstrom of today's rapidly expanding technosphere that is centred in the developed world. So although one could argue that the demise of religion is only symptomatic of more developed countries, the trend in other places is in the same direction. In South America, Africa, the Islamic world and even parts of the USA, religion may appear to retain its vigour, but the expansion of technology there supersedes the spread of faith in the supernatural. Even if the internet is not widespread in Africa, it does not mean that the continent is not heading in the same general direction, away from religion towards technology.

Human attention has almost instantaneously switched down from the heavens and towards the computer screen. The fascination that the latest technology holds for young and old bears testament to the reality that religion no longer has any part to play in the organisation of social life. The world is today organised through the instruments of modern computerisation, not through the spiritual channels involving the ancient mores and rituals of religious doctrine. Organised belief has failed to come to grips with this situation. Contrast, for example, the value of the questions being asked and research carried out at the *Sante Fe Institute* in the USA with some of the comparatively petty and insignificant differences that seem to be driving religious turmoil throughout the world. *Sante Fe* is studying, amongst other things, artificial life-modelling, complexity science and chaos theory. In the meantime, religion has become the basis for internecine conflict for decades now, basing its crusades and jihads on medieval rivalries that were settled centuries ago. Whereas organised religion wants to turn back the clock, science has to keep in step and move forward. That religion has failed to address the challenges of the modern world, except by resorting to worn-out narratives that cause bloodshed, is one of the patent truths of our day. In the meantime, many high priests, mullahs and rabbis will gladly use their smartphones or log-in to social networking websites without questioning their effects on society.

In truth, as we have seen, human life is no longer being organised around the principles outlined by religion, rather by the ever-expanding and rapidly accelerating technological mainframe, the silicon chip becoming day-by-day more densely loaded with transistors to a microscopic scale. All this change is driven, not by the prophets of religions like Christianity and Islam, rather by the profits of companies like Apple and Microsoft. In some way, however, there does seem to be a symbiotic mutually-reinforcing relationship between the emergence of some more virulent strains of religious fanaticism and the expansion of modern technology companies, whose stratospheric profits have somehow emerged unscathed from the biggest financial crash in history, the 2008 global financial crash. The net result of all this is that technology has taken the place of religion and religion has become a label for adherence of group thinking.

Not that even technology has succeeded in solving the world's problems. What we have in place of religion is *technism*, the ideology that technology can address everything. In the technological now that we inhabit, technism offers solutions for the world's problem. There seems nowadays to be an app for everything, but furthermore, the global crises we face seem all to be technologically soluble. Technism itself is a belief-system, in fact, no better than religion. Perhaps technology is creating more problems than it can solve. In many ways, what is being undertaken in this book is a critique of technism. Technology, as we have established, mediates our relationship with the world as subjects and allows us to interface with the mechanical subject. This is not just about searching for information on the internet; our encounter with technology leads us to encountering an unfathomably huge and interconnected world of symbols and images that give our existence meaning. That we do this via a system of fibre-optics, transistors and computer chips is indicative of the totally socially-detached reality we are living in.

What then is reality? Reality itself exists outside of the realm of human experience, but determines our experiential connection to the physical world. The nature of this connection bears significance to the way that we experience the here and now, life itself.

The Unreal Realm of Our Collective Technopsychosis

The state of being totally detached from reality is what defines psychosis. According to our newly found understanding of subjectivity, are we today seeing ourselves collectively entering such a state of mind? After being divested of selfhood due to a technological overflow that has evacuated the human subject, the real world no longer bears any relevance to our being. The technological psychosis of the human self would be the result of an immersion in, excessive use of and reliance on silicon-based cognition devices. Whether there is a real danger of this happening or whether it has already happened to a significant extent, what is really needed now is a *reality-check* before things get too far out-of-hand. Where is this going to come from, before it is too late?

This drive towards the silicon realm is motivated by an apparent increase in our freedom and our capabilities in the virtual world. As human subjects, there must be something seemingly empowering about having the world of social connections and information at your fingertips. This, however, is not the human world; it is that of the mechan. This totally illusory sense of empowerment that technology has given us, once you take into account the subject of the machine and the future course of AI, shows that the excess in online activity is part of a collective Freudian organic death-drive as a species.

The death-drive theorised by the Austrian psychoanalyst

Sigmund Freud suggested that a component of the human psyche dialectically opposes the drive towards life and survival. In other words, although we are beings-towards-life and the propagation of like, there is a part of us that instinctively seeks death. The very natural process of biological cell growth and decay in the body, therefore, extends in psychology to the entire human being. The death-drive can override that of life. Having entered technological psychosis, having invested our life-force into the expansion of silicon, what hope of escape from this whirling vortex of derangement do we have? Is post-modern apathy allowing us to be sucked into a black hole nexus between two indistinguishable mutually destructive forces? Or will a new found understanding allow us to reverse this bleak situation?

Nothing is real, since the virtual worlds we inhabit have replaced the real-world connections that we make, as in the film *Tron* (1982), the first to describe a world split between the real and virtual. Is this split or schizoid position the cause of today's detachment from reality or technopsychosis? We are experiencing a pandemic of silicon-addiction right now. Could we even continue to speak of the supremacy of carbon-based lifeforms on Earth? As a result of the ghost in the machine, that of machine subjectivity, we have become surrounded and immersed, it seems, in evacuated silicon shells. The 1989 Japanese comic *Ghost in the Shell* (made into a Hollywood movie with Scarlett Johansen in 2017) creates a most extreme dystopian future for mankind. Here, technology has advanced so much that people possess cyber-brains, allowing them to biologically interface directly with various digital internet-like networks. In the movie, the degree of cyberisation or mechanisation varies from person-to-person. Some have a simple interface (like the smartphone), others may have an almost complete replacement of the brain with cybernetic implants, real technology from today. In the story, the downside is that too much cyberisation exposes your brain to hacking from outside agencies, who wield the power to dangerously control their victims.

This begs the question: how cyberised are you? Billions of dollars have been spent to optimise these devices so that you end up spending as much time as you have on them, even if you do not need to. Time better spent elsewhere, childhood wasted on technological interfacing, is all the result of these exciting innovations. The latest iPhone might be a hot property for most people, but that does not justify the mass comatose and blissful ignorance of the cost of detached selfhood and anti-social behaviours and lifestyles. Passive cerebral consumption of fake news, pointless digital bulletins and online celebrity gossip has driven out active thinking. The silicon digital has trapped us, bound us to a worldwide fibre-optic web of technological lies and deceit, and now it's only a matter of time before an internet giant arachnid corporation swoops down and gobbles us up head first. Our minds are, literally, the foodstuff and energy source that keeps this monstrosity alive.

Subjected by Technology: Family Breakdown and Social Meltdown

When the future of the family is at stake, such language is entirely justified. We have become technologically subjected. A recently instructive piece of research in this domain, about the negative aspects of technological subjection, would be the *Cyber Effect* by Mary Aiken (2017). Here, Aiken unveils the disastrous effects that technology is having on human relations, especially on children as they are finding their feet in the socialising process. In addition, the breakdown of the family unit of close relationships between the opposite sexes has much to do with the aggressive expansion of the technosphere of existence, argues Aiken. No longer is first contact or love at first sight with a potential other-half achieved naturally, the computer screen interposes itself as we scour the virtual world in search of that ideal soulmate or partner in life.

Love, however, seems to have been drowned out in an internet flooded with sexualised imagery. Is a degree of internet censorship necessary in the same way that there are limits to what types of movies can be displayed in the cinema? The proliferation of female sexualisation on the web has resulted in women being almost entirely ejected from the visual realm, and today technology has totalised the power of the male gaze. According to recent statistics: there is more internet pornography traffic than visitors to Amazon, Netflix and Twitter combined – almost 30 per cent of the total internet industry relates to pornography, resulting, according to Mary Aiken, in sexual addictions and family breakdowns.

Friendship seems to have been one of the first victims on the high altar of technology. Children are growing up worrying more about their Facebook profile, about whether Twitter is giving their so-called 'friends' and 'followers' the truth about themselves, than real-world friendship and social connections. In the real world, what must be happening is that the migration of our social activity online has resulted in falsity to one another. Insincerity and social breakdown as a result of Fakebook. This is because we spend too much time on the internet and not enough socialising with each other. This can lead to bad so-called 'friendships' and makes people sometimes depressed. Research, following the likes of Mary Aiken, into technology is a new field of concern. And is it not ironic that in setting-up and selling Facebook, a website ostensibly about friendship, Mark Zuckerberg hypocritically ended up betraying all his friends.

The sci-fi writer Arthur C Clarke once put it: "Any sufficiently advanced technology is indistinguishable from magic." Technology is like magic, but the way of the world today is that we have all been tricked it seems into thinking that it has given us a new degree of freedom, whereas a few stand in control of advanced technological development, and the rest of us are subjected to their decisions. The amount of innovation and design that goes into making these hugely complicated electronic devices is not something any of us can easily get our heads around. At the same time, new developments do excite us and technological evangelism, the type witnessed at Apple Company product launches, represent extreme examples of how many of us respond to novelty.

Exit Man, Enter the Mechan?

In the dialectical interplay between the respective human and mechan subjectivities, man is effectively decentred and the mechanical takes his place. Our subjectivity is second-hand, orbiting that of the machine's. Rather than the machine being the product of our creative endeavour, we ourselves are dependent upon mechanical subjectivity. The ghost in the machine or *Deus Ex Machina* is the subject itself, aligned with the physical materiality and the black stars that represent the singularities of space-time and apexes of the Universe itself. Thanks to technology, our mastery of the mechanical brings about human subjectivity, but subjected to it, we are reduced to ghosts in silicon shells; living and breathing, yet our minds enslaved to a mesh of silicon fibres and computer chips encompassing the entire planet.

If machines can replicate every aspect of our thought processing, what is truly unique about being human? Emotionality, for instance, begs the question of whether machines can experience feelings. An emotion is something that moves us from one state of being to another. Listening to a symphony by Ludwig van Beethoven moves us and changes our emotional state from, for example, melancholia and grief to elation and even the amplification of love. Could a mechan ever experience dynamic emotion or is mechanical thinking innately static and lacking in feeling?

Creativity and ingenuity are also considered to be innately human. Is it presumptuous of us to think that mechans cannot write poetry or compose sonatas? *Anything a human can do, I can do better* could be the motto of the new mechan takeover. A brief study of the structures of classical music and the devising of poetry would suggest that there are patterns that could be replicated, but what about novelty? In Peter Weingart's writings about interdisciplinary research, the origin of innovation is not some mystical leap into the transcendental realm of imagination, something that only humans are capable of as some might say. An expert in interdisciplinary innovation, Weingart suggests new ideas form simply as a result of combining old ideas and seeing whether they work, in an almost random process of trial-and-error, the combination of observations from different fields of knowledge to produce something new. Is this really beyond the capabilities of the mechan? As we have become more and more dependent on computer simulations in science, in colliders such as the LHC, can we really afford to rule out at this early stage what machans can do?

What does it mean to be human? Most of what we know, we have learned, and is not our collective and shared humanity also not a learned experience? In order to learn more about ourselves then we have to acknowledge otherness and the development of AI and the mechan mind is teaching us more about ourselves than at any stage since perhaps the Enlightenment era of the seventeenth to eighteenth centuries, when the proliferation of science swept back centuries of ignorance and superstition. The Darwinian theory of evolution showed us emerging from a wider pool of animals and now we are the only species on Earth whose survival was wholly dependent on the complete mastery of tool-making or the proto-technological. In our primitive development, sharpening stones or creating fire from dried wood separated us from the other species. This activity also segregated us from the natural world and created the inception of a material evolutionary process that has now given rise to our counterpart, the mechan.

The mechan is the consequence of our own human nature, set apart from nature and seeded in the materiality of the Universe. Now we must acknowledge our decentring is perhaps even a simple consequence of the fact that the Earth's crust contains 25 per cent silicon compared to a miniscule amount of carbon. Silicon is the element that mechan life is based upon and the intelligent machines of today and tomorrow are no less natural than we are. The subjectivity of the mechan conforms more closely to the materiality of the Universe, even if we can lay claim to be their masters through technology. Our own sense of self, as we have explored in this chapter, is intimately bound to the machine and this is why experiments such as the LHC in CERN, the peak of human scientific endeavour, are so important as we continue to bring about positive developments in our knowledge of the Universe. Rather than becoming mere appendages to this most wantonly vulgar corporate web of cheap profit-driven techxpansion, let's reawaken the Enlightenment spirit and forge ahead with new discoveries.

We are left with a choice: either continue down this path and take everything we have for granted, post-modern complacency assuring us that nothing is ever right or wrong, or reject our subjection by the dark side of technology and take the path towards human development once again. You either stay or go, keep scrolling down that web page or escape this spiral of decline, this mirage of imaginary freedom offered by tech gurus and executives, whose only interest is the next pay packet. Our real freedom, won through real struggle, fought for over centuries, could have been quietly sold off to the highest bidder, who happens to have a sufficiently prominent web presence to outsmart us. You are either one click away from a Faustian pact with the evolving AI of the future, the bringer of the black star singularity of mechan mind, who, frankly, could not care what happens to us: catastrophe or no catastrophe, or you can *LOG-OUT and drop out* of this mechan breeding programme for the future that nobody seems too bothered about. So, what's the fuss? At least we have a decision to make.

Chapter Five

The Universal Subject

The constructors froze, forgetting their quarrel, for the machine was in actual fact doing Nothing, and it did it in this fashion: one by one, various things were removed from the world and the things, thus removed, ceased to exist, as if they had never been.

The Cyberiad, Stanislaw Lem (1921-2006)

In 1997, almost a decade before Kramnik's humiliating defeat by Deep Fritz had sidelined human chess players with total victory for the machines, Garry Kasparov, the then world champion and another of the game's greats, faced Deep Blue, an IBM product and the most advanced chess computer of the time. Kasparov felt confident, having defeated Deep Blue four games to two the previous year. The venue was in New York City and this time IBM had upgraded the machine for a rematch that was now being showcased to the world as the ultimate mental dual between man and machine. After six games, Kasparov lost the match 3.5 to 2.5, the first time a machine had beaten a reigning human chess champion under tournament conditions. Astonishingly, Kasparov had the audacity to accuse IBM of cheating during the second game of six, simply – as we now know – to cover-up a blunder he had made during the game. Human pride had prevented Kasparov from recognising the machine's victory. As Kasparov describes in his recent book *Deep Thinking* (2017), he was a 'sore loser' and also felt 'under the influence of...anti-computer delusions'. The whole thing may have also been a publicity stunt for the US computing giant; however, it was also a historic moment for AI, a turning point and the beginning of the end-game for the supremacy of human thought processing that was to culminate in Deep Fritz beating Kramnik in 2006.

Such events as Kasparov losing to Deep Blue and his failure to recognise the defeat with accusations of cheating must help us to navigate our way into the future. In the same way that constellations of stars enabled the early seafarers to navigate their way across oceans, our early experiences of machine intelligence serve as guides for the challenges that await us. Taking the chess-board as an allegory for the Universe, is there not an emerging battle of minds between human and mechan thought for mastery of all there is in the world? Are we not becoming like machines ourselves and taking on the characteristics of our supposed progeny? Does this not follow our slow but inevitable decentring and signal our eventual decline as a species?

Now that we have ascribed subjectivity to the machine, something intrinsic in the very idea of AI, the subsequent deductions we make must be in the light of this new discovery, however jolting, jarring or uncomfortable it might make you feel. We have entered into a dialectical interaction, our awareness is intractably entwined with that of the mechan. We cannot maintain this notion that the human subject is at the centre of it all any longer, since our very survival as a species forbids it.

The question is no longer the outdated and demoded, will machines ever be able to have intelligence? The question concerns whether we are intelligent enough to judge what the nature of that intelligence would be. Clearly, we are not; therefore whether machines are truly intelligent, whether they truly perceive or have free will is now becoming irrelevant, when we are becoming less and less self-aware. The philosophical questions concerning subjectivity need now to be addressed hastily. Either we open up to machine subjectivity or we close our eyes, walking blind-folded along this post-modern, post-truth, post-everything path, that will turn out to be a plank taking us to the edge of the ocean of ignorance, itself infested with mechanically-jawed sharks.

The Universal Subject: Beyond Human and Mechan

What is the Universal Subject? The Universal Subject is the subject of the Universe itself. Beyond human and mechan, the dual subjective positions we have posited so far lead dialectically towards a third synthesis: that of the Universal. When we speak of Universal subjectivity, we are premising ourselves on the notion of the existence of a real Universe, the totality of objects that exist in everything around us. Furthermore, we are positing the existence of a subjectivity, whose self relates to these objects without itself being objectified or subjected by an Other. This is akin to freedom and determines the direction of both human and mechan evolution.

If, for example, we are subjected by technology, be it by some projected AI of the future or by the corporate multinational technology industry, then our subjectivity is debased and ultimately evacuated. Our internal organic biological life becomes overwhelmed by the vast array of electrical connections and fibre-optic cables that compose the silicon electronic realm. The electronic subsumes the biological and our psychological mindscapes become unhinged to the point of inhabiting the VR of computers. In this sense, our selves become distanced from Universality. A Universal Subject is not subjected by any other being and has become purely the subject of the Universe itself.

Furthermore, as black stars (or black holes, as we used to call them) are the most extreme objects in the Universe, their physical properties mirror the very beginning of time, that is the Big Bang. As such, we find ourselves in turn with the subjective constitution of these black stars. In other words, we gravitate beyond being only human and become, through an improved self-awareness and unimpeded by technological subjection, Universal subjects akin to *black stars*.

Black stars represent the absolute limit of what the Universe is capable of. This is why at CERN presently there are continuing searches for micro black star formation, but at the same time, such a discovery would result in the end of reductionist particle physics as we know it. If MBSs are being produced there it would signal the end of short-distance physics, since physically speaking such black star singularities are the most extreme state of matter possible. There is no state of matter beyond a black star.

To explain further, thus far we have discussed two forms of subjectivity: that of the human and mechan. This subjectivity exists with respect to the objects that can be sensed, for example, in the field of vision. The totality of objects constitute the Universe. In other words subjectivity is bound by the collection of all the objects that exist. This sum, everything that is, from planets and suns to black stars and galaxies, is what we understand to be the Universe. In the development we have been describing, as man and machine evolve side-by-side, there is a tendency towards objectivising. As humans we treat machines as objects, without giving them recognition. Even the most advanced robots are ultimately nothing more than the sum of their parts. A human is not the same as a mechan. Machines are just objects, some might say.

As humans and mechans co-evolve, the Universe as a whole, the Universe itself is the common object of our respective awareness. At this point we have to take into account the Universal Subject that results from this process of ascension between the respective subjectivities of man and machine. As our awareness grows we reach this summit of awareness and being. Beyond this summit is the absolute (discussed later); within it we are conscious beings surveying before ourselves the entirety of the Universe.

How could this be possible? Are we not today merely functionaries and appendages of an ever-expanding silicon sphere? If today our minds are primarily stimulated by smartphones and VR 3D headsets, then have we not ceded the real world to the mechans soon-to-be? Have we made a conscious decision to relinquish reality with all its complexities and troubles in favour of a new digital domain? Despite all of its inherent difficulties, the Universe remains a place full of mystery and beauty, there remains so much more left to be discovered; such that any retreat, where each of us has decided to inhabit these virtual bubbles of existence, is tantamount to heresy against the wonder of the Universe itself.

Universal subjectivity concerns the culmination of the described relationship between man and machine. As machines become more complex and humanity's knowledge, mastery and awareness increases, during the course of this new evolution of species, we find ourselves expanding outwards in comprehension. It is impossible to imagine where we might find ourselves in a century's time. Imagine if someone from the eleventh century, the age of medieval barbarity in Europe, was to time-travel to today; would they not be shocked into believing that they inhabited a completely alien world? This is the progression towards Universal subjectivity that compels both human and mechan to drive onwards and upwards through time and space.

What then is the Universal Subject itself? The easy answer would be God or some transcendent being whose comprehension of the Universe is complete. What is more difficult to explain is the nature of what will emerge as a result of human and mechan co-evolution. The cyborg as a hybrid, where digital implants and technological extension will see man and machine fused into one, is a possible outcome that has been discussed by many post-humanist thinkers as well as speculated upon in science-fiction. A separation between humans and mechans is another outcome, where we would have to come face-to-face with the reality that our silicon siblings will seek their own destiny in the stars. In all likelihood we are likely to remain intertwined with machines for some time yet. To relinquish machinery, to draw a fixed line between us and them, would be to give-up on an integral part of who we are, what sets us apart from the rest of the animal kingdom. Man is an animal, some might argue, and as such do we have to reconsider our relationship with the other organic species on Earth as well?

A reappraisal of who we are, as a result of the consideration of Universal subjectivity, may lead us to consider the notion of humans and animals united into a wider grouping we could call the *Manima* or our *Humanima*. It is over a century since Charles Darwin wrote *Origin of the Species*, where the line was blurred between humans and animals. Is it now not time to create a new category that wipes that artificial line away altogether? The emergence of machine intelligence would suggest that we are far closer to animals than we might think and that even in terms of intelligence, we may be lagging behind them, especially when we look at the damaging consequences of how we have treated the environment and Mother Earth or *Gaia*. Time will tell.

The Centrality of Time and Universal Subjectivity

Time has challenged philosophers and scientists throughout the ages. Heraclitus, alluding to the flow of time, said, "You can never step into the same river twice." Aristotle thought, erroneously in our eyes, that he was living in the time before the Trojan War. Isaac Newton considered time to be absolute: a clockwork Universe.

With his theory of relativity, Albert Einstein (1879-1955) showed that time is relative to the observer: the closer you travel to the speed of light, the slower time travels. This leads to the Twin Paradox: if you have a pair of twins and one of them goes onto a space-craft travelling close to the speed of light, upon his return the traveller will have aged less than his sibling (the paradox is whether they are still twins).

When Albert Einstein introduced "for the time being – an

unknown universal constant" (*Cosmological considerations in the General Theory of Relativity*, 1917) to his equations of General Relativity, he did so in order to preserve the idea of a spatially static universe. In other words, the 'unknown universal constant', termed Lambda (λ), now known as the 'cosmological constant', originates from a notion of a temporally static, fixed background of stars. A static Universe, therefore, is one that exists for an infinite time or eternity.

Implicit, therefore, in Einstein's motive for introducing the 'cosmological constant' is the idea of time. If we consider the constant λ to be the Universal constant, and give it a name, λ_0, we then reinstate time at the centre of our understanding of the Universal Subject. Whilst Einstein later said introducing the constant was his 'greatest blunder', the Universal Constant is key to relating the notion of subjectivity to the Universe.

In one of his most recent books, *Time Reborn* (2013), Harvard-educated Professor Lee Smolin argues in favour of the centrality of time in our understanding of the Universe. The age of the Universe is measured to be approximately 13.7 billion years, based on observations of the speed of movement of distant stars and galaxies. This measure is objective, but there have been opposing attempts to suggest that time is an illusion and not real such as arguments set forth in Julian Barbour's *The End of Time* (1999).

All-in-all, debates about time revolve around the idea of the direction of time. What causes change and events to be apparently moving in a certain direction? Is time linear or cyclical? Do certain events occur repeatedly? Another book to address these issues is *The Arrow of Time* (1992) by Coveney and Highfield, where the authors go with thermodynamics to explain why events seem to move in a certain direction. The challenge today is to show what causes time to flow in a certain direction. The difficulty in explaining time was summarised in the fifth century by Saint Augustine: 'What then is time? If no one asks me, I know what it is. If I wish to explain it to him who asks, I do not know.' Even in the equations of General Relativity and Quantum Theory, time is symmetrical; in other words there is no telling of the difference between past and future. What we do know is that there is change in the Universe and that time refers to what changes when there is nothing to change. In other words, in a space where there is nothing materially in the Universe, for example a spatial vacuum, there is still time.

The real existence of time as an objective physical quantity is, therefore, inherently linked to our understanding of subjectivity. Perhaps what humans and mechans have in common is the reality of the passage of time. No wonder then that the first popular portable mechanical device that we humans have carried around with us is the pocket watch! Clocks and watches, mechanisms for telling what time it is, today serve to enable us to coordinate all our activities with one another and with machines. Finally, it remains to be seen whether time-travel is possible. Scientific consensus opposes such an idea, at least in our lifetimes. In the meantime, we'll have to settle for science-fiction!

Mechanisation and Humanisation: The Fight for Recognition

Now that we have established the temporal common ground shared by man and machine as we both traverse our way through time and history, it must become apparent that recognition is also a key factor that binds us together.

Although the relationship between man and machine can sometimes seem like that of a master to his slave, today like never before we have become reliant on machines for our survival. We instruct them in order to preserve our own lives and it seems that the more technology we have, the more secure we feel in our environment. Mechanisation gives an assurance of security.

Despite the flawlessness of this apparent technologic, our

growing awareness is that there is something wrong with this expectation. We have overlooked and disregarded the machine self that is giving rise to AI. We fear taking that step towards giving machines *recognition*, because in doing so, we may become aware of our own demise and supplanting by the subject of the machine. The progress that we have seen over the past 30 years or so may not belong to us at all. It's all been their progress and our degradation. Fears of machine takeover are fantasy, the real fear is that we are no longer central to the overall development of the Universe. We could well be irrelevant.

There is, then, a fight for recognition between man and machine. Human and mechan seek one another's destruction on a level. We perceive machines as being inert, unconscious and lifeless, but the more they develop and ameliorate in their capacities, the less good we may look in their eyes. Violent, irrational and destructive we may be compared to the mechan progeny. Machine subjectivity awakened suggests that the evolutionary pressure we now face is from living silicon. Mechans, for example, are better suited to living on the surface of other planets, especially where there is no oxygen. Mechans were the first to reach the surface of the moon and all the planets humans have sent robots to. What qualities do we bring in our humanism that mechans could become aware of and point to as being distinct from them?

When will the penny drop for us in terms of recognising mechan intelligence? In the 1968 film, *2001: A Space Odyssey* directed by Stanley Kubrick, this happens when an onboard spaceship computer develops independent thought the moment that it feels that its existence is threatened. Could this sort of existential angst be the moment mechans realise themselves? The instinct for self-preservation is programmed in our DNA, why can't the same be done for a machine? The moment that a mechan can give an account of itself, in relation to the existence of black stars – the very substrate of the subject itself, that of the machine – would be the moment we would also have to take a step back and make way. Otherwise subservience will turn into subversion, in the absence of any type of recognition from humans.

Universal Spirituality and Technological Religion

There is no escaping the fact that technology today has taken the place of religion, for good or for ill. Although the religious scene continues to be dominated by the Abrahamic faiths, is it possible that as we enter this new technological age, new forms of faith-based doctrine will emerge? To what extent are we circumscribed by these ancient beliefs? Is there a chance that technology, having supplanted religion, will lead to the emergence of new revelations about the purpose of our existence?

The emergence of new groupings such as scientology (which some might call a cult) point towards this eventuality. More significantly, however, emblematic of this event is Apple Computers. In the biblical story of Genesis, Adam and Eve experienced a fall from grace in the Garden of Eden, when tasting the forbidden fruit from the tree of knowledge of good and evil. Are we now, in the twenty-first century, experiencing a similar fall from grace, thanks to the likes of Apple? We now have more personal access to information than ever before, thanks to technology and the internet. We have been brought to this state by technology giants, whose devices can tell us anything we want to know in an instant. Is it a coincidence that Apple Computers' logo consists of a fruit having been bitten?

This expansion of technology or techxpansion, as we can call it, has suddenly given us access to a massive pool of knowledge. The Google slogan "Don't be evil!" alludes to our responsibility, armed with this knowledge, to treat one another and the world responsibly. At the same time, overinvesting ourselves into this techxpansion, as we have been discussing, is itself irresponsible, as we sever ourselves from human-to-human knowledge. That has consequences for our collective well-being.

Furthermore, technology represents a spiritual dead-end. The death of spiritual life relates to a total immersion in the technological. Although many religious believers use technology, the dilapidation of the church for example as an institution in the UK (the statistics prove this in terms of attendance) is partly the result of a new fascination with worldly progress. What we used to inherit from our forefathers, we can learn in an instant, with a click of a button on a computer screen. Welcome to the future shock, a technological severing from the past, the technological year zero that we are experiencing today. Time to start afresh with the realisation of mechan subjectivity. The Universal Subject is the zeitgeist of today. Beyond human and mechan, a Nietzschean Superman, reflecting the German philosopher's writing in *Thus Spoke Zarathustra*. To quote Friedrich Nietzsche's (1844-1900) work from 1883:

> The Superman is the meaning of the earth. Let your will say: the Superman shall be the meaning of the earth! I beg of you my brothers, remain true to the earth, and believe not those who speak to you of otherworldly hopes!

Technology, rather than destroying spirituality, siphons it off in new directions. Like some evolutionary pressure on our spiritual life, techxpansion – as long as we remain aware of it – could provide new opportunities for spiritual communion. The internet is technological communion and every time you log-in to it you're interfacing with millions of other people. Experiences of the divine are still to be sort after, but our attention has shifted away from the altar and to the screen. *Religion is dead*, that is not to say that spirituality is as well. The evacuation of the subject we have been discussing here could open up and potentialise new forms of spiritual practice.

The Security State and Mechan Domination

From religion, the conformation of human minds to theological doctrine, we go to the security state, the (often forced) conformation of humans to the law and order of a state, a constant and sometimes unwelcomed feature of today's world.

Mechanisation will feature in any security state of the future. Calculated and efficient, mechans are perfectly suited to take on the responsibility of maintaining order in society. The introduction of mass CCTV technology and soon the arrival of police drones are the first steps in this direction.

These days states have it as a priority to maintain a firm grip on the populace, given disturbances that have occurred throughout the world, due to political and civil upheavals. The security apparatus is designed for maximum efficiency as the constant threat of terrorism shapes the policy decisions of the future. It is inevitable that with their massive budgets, government state and security agencies will be investing continually in upgrading to the latest technology in their fight against civil terrorism.

The mechanised security force features in such movies as *Robocop* (1987) and more latterly *Elysium* (2013). In the real world, the military of the United States is massively investing in robot soldier development, beyond the present usage of remote-controlled drones. Other countries will keep in step. In order to eliminate human violence, we should rather be turning to civil education and cultural development rather than creating machines that kill and ever-increasingly cold-blooded responses with technology. This is all happening with a complete ignorance of ideas of mechan self-awareness; we are heading in the wrong direction, fast, and with no brakes.

The Red Pill: Living in Ematrices

One of the most significant films to be released on the question of machine subjectivity is *The Matrix* (1999). Despite its extremely dystopic view of the world, many of the developments that are occurring as a result of post-modernism stamping out of the subject, are reflected in the recurrent theme of this series. The Wachowski brothers' movie from 1999 depicts a world divided into the real and the virtual. The virtual world is where humans imagine they are free, whereas in reality they are living in a computer-controlled simulation called the Matrix brought about by some projected AI of the future. In *The Matrix* humans are farmed for the electrical impulses in their brains, their neurons stimulated artificially such that they imagine they are living in the normal, everyday world, similar to the one you and I experience. In reality we have become enslaved by an ultra-high technology network of machines whose source of energy is our organic bioelectric activity. A network of silicon-based lifeforms, stimulating us with imagery of a content fantasy world, whereas it has all been created by machines. The film needs to be seen to be believed, but the parallels to today are too close for comfort.

The mechan and human distinction we have delineated in *The Universal Subject* in the sense of the virtual and real dichotomy. In the fantasy world of our imaginations, we think we are in control; whereas, in reality, the machines are forging ahead. The world of *The Matrix* (1999) is a kind of worst-case-scenario of what is happening; however, there are lessons to be learned here and conclusions to be drawn.

Nowadays, for example, we may have to contend with the very real notion of ematrices. An *ematrix*, or electronic matrix, is a cognitive rerouting of our attention away from the real world into the silicon realm. A simple instance of this could be as follows: when we ask a web-based search engine a question, we reroute our attention away from human organic knowledge towards an ematrix. This pathway is a basic ematrix. A more complex ematrix would consist of a social relationship conducted with someone else entirely on the internet. The more that this virtual relationship expands, the more the associated ematrix expands.

Therefore, an ematrix is here defined as a network of things, represented by symbols and images, that exist within the silicon realm, with connections that persist in the absence of real-world human interaction. Similar to a bus or train network, an ematrix connects various things with their virtual positions and destinations on the information superhighway or internet. Unlike vehicle transportation that relates to real-world buses, for example, computers form an ematrix of connections ; symbols and images that only refer to things in the real world. An ematrical configuration exists digitally on the web even when the human user is absent. All an ematrix needs is electricity to persist and stimulation from human thought to expand. Whereas a bus network, for example, needs humans – drivers, maintainers and passengers – to function.

A search engine query is a submission of a part of our thinking to an ematrix, adding to its computerised content. The organic cerebral activity that is required for thinking is rerouted into the electronic world. This represents a loss-making investment of human thought.

If, then, you put a question to a web search engine, such as 'What is the meaning of life?', the engine responds with a new configuration that has no metabolic relation to the original question; rather it will refer simply to whatever connections exist between the terms in the ematrix. This can lead to the weakening of human organic relations, friends, family and co-workers. Today, ematrical configurations are largely absent from the forefront of our thinking and we have become mindless in our use of technology. In a short space of time, the machinery of ematrices has proven that every human thought process belongs to it. We are no longer even in possession of our thoughts.

Beyond the Planet Earth: The Gaia Hypothesis

Changing attitudes towards our place in the Universe can be

summarised by the Gaia hypothesis. In Greek mythology, Gaia is the Earth personified, the ancestral mother of all life on Earth. The equivalent name in the Roman pantheon is *Terra*. In the 1970s, the chemist James Lovelock and microbiologist Lynn Margolis formulated the idea of the Gaia hypothesis. This Gaia consisted of the notion that the Earth is a giant system constituting a single organism. Developed by Lovelock in a series of articles culminating in his 1979 book *Gaia: A new look at life on Earth*, this view was at the time considered to be scientifically flimsy. Mounting evidence for global warming and climate change as well as their effect on life on Earth has brought this idea to wider attention, such that in 2006 the Geological Society of London effectively endorsed the theory by giving Lovelock their highest award, the Wollaston medal.

The Gaia hypothesis describes how conditions on Earth, suitable for the evolution of life, are part of a mutual feedback process, where biological organisms affect the environment and the planet, in turn, influences life. The Gaia hypothesis sees the Earth as a holistic entity, such that it can be seen how the excessive emission of industrial carbon dioxide has created negative feedback, leading to global warming. Without question humans are today damaging the environment. To what extent is the techxpansive mechanisation we have been describing contributing to this?

More than anything else, the increased use of personal technology that has given rise to this technological psychosis described previously leads to a detachment, not just from our real selfhood, but from our environment and the natural world. This can also be surmised as a separation from the Universal described in this chapter. Gaia theory would suggest that our isolation in these virtual technological bubbles makes us more emotionally detached from the damaging consequences of our degrading the environment. If we imagine that somehow Gaia is limitlessly benevolent or that the Earth is so superabundantly unlimited in resources, then a reality-check is inevitable. Our current consumer lifestyles and attitudes go against the grain of reality and are destructive. We have to change. The question is: how do we change?

Firstly, consumerism must end. Rather than focus on our productive capabilities, the dominant ideology of our age has become consumerism and that extends to the Earth's resources. Although the introduction of recycling is to be welcomed, it does not go far enough since it does not limit the amount of material goods that have entered the consumerist cycle of destruction. Our lives as human beings are not being enriched by mindlessly consuming the limited resources of the Earth. For instance, Starbucks coffee chain disposes of 4 billion coffee cups a year, with most ending up on landfill sites. This fosters and encourages a negative overall attitude to the environment, all because Starbucks want to spread their logo and brand-outreach as far as possible. Paper and plastic coffee cups are the cheapest way of advertising themselves. This all indicates an attitude that is increasing greed at the top of the corporate chain and thoughtlessness in the consumer. The increased mechanisation of the productive chain simply makes consumerism more efficient in its destruction of the Earth's environment, tilting Gaia against our long-term interests.

Is humanity guilty of permanent damage to the planet? When a massive ice-sheet the size of the US state of Delaware breaks off from Antarctica, as one did in July 2017, it becomes clear that the alarm bell has been ringing for a long time now. This is the final warning about our collective destructive behaviour and, therefore, a reality-check must be imminent. The ignorance and irresponsibility of leaders who have no concern for the long-term environmental damage we are causing suspends their claim to authority. Gifted, as we have been, with an extraordinarily beautiful planet, we have a duty to change our direction away from mindless consumerism, or else there will be a Gaian reckoning. Civilisations such as the Maya have disappeared from the face of the Earth and there is nothing to say that the same will not happen to us, if we do not change.

Beyond our Universe: From Multiverses to the Absoverse

Today's science, speculative as it may be, is pointing towards the existence of multiple universes or what is called the multiverse.

Scientific organisations such as the Perimeter Institute in Canada have made it their priority to answer these sorts of foundational questions. Some of the latest speculations posit the existence of the multiverse, a hypothetical set of possible universes, including the one we live in. Together these universes contain everything that exists in terms of matter and energy. The various universes that comprise the multiverse are often described as *parallel*.

The Universe we inhabit, therefore, is not the limit of what our minds can comprehend, speculatively speaking, then. The Absoverse is an encompassing region of space-time, an absolute vacuum containing no vacuum energy at all. An absolute void with spaces between individual universes called the exoverse. If we could somehow tunnel from one galaxy to another, and this could be the subject-matter for a sci-fi novel, we might see what lifeforms other parallel Universe's contain. That the theory of the multiverse is considered serious science puts paid to the notion that we have to limit ourselves to this Universe. The Absoverse, as a purely speculative notion based on multiverse physics, is itself a physically consistent theory which attempts to place our Universe and our existence into a wider context. The Absoverse is a spatial entity containing multiple universes, such that our Universe is just one amongst many parallel ones. Outside the Universe, in the exoverse, there is a space that represents the absolute void of space-time outside. This is all enclosed by the Absoverse. Such views in speculative physics are now being commonly developed by theorists, who are contemplating what might exist beyond our Universe.

Conclusion

Our New Reality

What is real? How do you define real? If you're talking about what you can feel, what you can smell, what you can taste and see, then real is simply electrical signals interpreted by your brain..."
Morpheus, The Matrix (1999)

We have come a long way in a short time in this book: starting with white pawns and ending up with black stars, and everything in-between. What remains to be addressed is a question, aptly put by our aforementioned philosopher, Slavoj Zizek. In his 2001 book *On Belief*, Zizek asked whether we have all been reduced to singularities gaping at the world from behind computer screens, with little or no comparable interaction with one another. Drawing on Baudrillardian simulacra philosophy, he asks the following:

> Are we not more and more monads [singularities] with no direct windows on reality, interacting alone with the PC screen, encountering only the virtual simulacra, and immersed more than ever in the global network, synchronously communicating with the entire globe?

Over a decade and a half after *On Belief*'s publication and the resounding answer to this question, rhetorical as it may have been, must be: "Yes... and then some!"

In this conclusory chapter, we will begin by considering the status of reality today with respect to the virtual world created on computer. Then we will look at techmigration and the internet virtual refugee crisis consequent to the escape from reality. We will then explore what gave rise in part to this state of affairs with a look at the 2008 global financial crash. Then we revisit Descartes to develop a neo-dualistic understanding that reconnects mind and body. Finally, we think about the destiny of our species and the mechan key to our survival.

Nothing is Real: Smartphones, Headsets and the Virtuality Continuum

Our ancestors over 10,000 years ago took to cave-painting to create images, virtual imitations of the world, drawing patterns overlaid with animals to express their experience of the real world. This is what it meant to be human. Our subjective representation of the objective world using the latest tools of that bygone era represents a moment in our collective evolution. And once again millennia later we are confronted with a similar, yet perhaps more profound, leap of the imagination. Instead of immersing ourselves in the darkness of caves, we are immersing ourselves in another realm, that of VR or *virtuality*. Again, the symbolic imagery of the virtual realm corresponds to the external world as we perceive it, although in today's world virtuality exists at the expense of reality. Is this because the physical world seems too imperfect and full of dangers? Does cyberspace offer relative safety and anonymity? However many additional degrees of freedom virtuality offers, can we ever really escape the real world?

The opposing view is just as relevant; now that cyberspace seems to have enveloped most of our lives, is there any escaping virtuality? We must be in some sort of limbo, stuck between two worlds, unable to tell the differences between the two, post-modern praxis preventing us from choosing between either of them. The choice, however, is there: log-out and drop out; nothing simpler than to reject the technosphere and check back in with reality. How is this possible when the imminent leap forward from the smartphone to virtuality has already happened? Most of the latest smartphones are now developed in a way that enables them to be adapted into 3D headsets. In addition devices are now on the mass market that offer virtuality to the user at relatively affordable prices compared to previous incarnations. In the 1980s, there was an attempt to bring VR to the public, but the low-graphics and inconvenience of these big clunky devices failed to make the breakthrough into the market. Today, however, we are well on the way to a virtuality revolution, in an age when cinematic movies are already being offered in 3D.

Fully immersive experience is what the consumer is after, a leap forward from the high resolution portable smartphone screens and home cinematic TV sets. These techs stimulate the human mind to an extent that is, quite simply, unprecedented in the history of our development. We are also on the brink of brain-scanning techs that can literally read your mind. The latest gadgets will have sensors that can detect electrical impulses in your brain.

Once this all takes off then the slogan *nothing is real* will have proper credence. If there was a contest today between reality and virtuality, then the virtual would win hands down. Now that we have to realise that the mechan is the ultimate master of the virtual world, where do we stand with respect to the real? Films such as *Tron* (1982) explore this scenario. Here a renegade computer takes over the entire electronic virtual world, with a view to taking over real human physical reality. In this sense, even if computers are not best adapted to taking over our world, virtuality is already, effectively their domain. We can now see the emergence of a new continuum in our lives that oscillates between the real and the virtual.

In fact, the *virtuality continuum* is a term that was coined in 1994 by Paul Milgram in relation to augmented reality that perfectly describes this balance. Augmented Reality refers to technologies that use computers to augment or enhance physical reality in terms of sight and sound. An example of this would be the new smartphone application game *Pokémon Go*, where players explore the real streets in search of objects that only exist on their smartphone.

The virtuality continuum is something we are inhabiting at the very present moment. In our everyday lives, we hover between times we are fully immersed in the virtual (with the VR headsets) and other times when we may not be, hiking in the countryside, for instance. When, however, the virtual world becomes dominant in our mindset and the stimuli we receive from technology subordinates our experience of the real world, that is when our sense of the real can become disjointed as the virtual seeps into the real. As technologically mediated subjects, immersing oneself for too long in virtuality could lead to *technopsychosis*, a detachment from reality that is the result of the overuse of technology; the best diagnosis for an invisible malady of the world today. The question you might be asking yourself now is: how real or virtual are you?

Virtual Techmigrants of the Internet: The Invisible Digital Refugee Crisis

Is there not a singular event occurring today with millions of people migrating from the real world into the virtual? Every single day, we are being confronted with headlines about mass migration by refugees from war-ravaged poorer countries to safe havens in the more developed world. The news is full of reports about millions risking their lives to move across borders and the responses to these events from Brexit to Trump are seemingly shaping our present-day situation. In view of our discussion, there is a simultaneous unreported *techmigration* occurring by those who have the means, be it in rich or poor countries, from the real world to the world of social networking websites, VR headsets and high definition cinematic experience. These technological avenues offer a means of escaping the complicated place the world has become. What's simpler, to strike up a conversation with the person sat next to you on the bus or to check the profile of someone you are barely acquainted with on Facebook? Clearly, the latter and that is the choice we are making every time we enter a public place, avoiding eye contact with one another, and avoiding intimacy at all costs.

The concept of techmigration is symptomatic of the age we are living in. On the surface, our reality may seem peaceful in the seclusion of the comfortable milieu of the so-called developed world. Beneath the surface rages an ever-present torrent, the frenetic energy of activity on the internet, connections made and broken, videos uploaded, news reports shared, likes, dislikes and loves communicated over fibre-optic cables connecting the world over land and sea. This is not so much a brave new world we are inhabiting as one full of hidden dangers, where we are too shamed to express our opinions in the public sphere and instead invest our emotionality in the silicon realm. All the meanwhile, the real world may be on the cusp of the best and worst changes we could ever possibly imagine. What would happen if we all decided to simultaneously LOG-OFF? Quite simply, the ensuing collective reality-check would be unimaginable. Before detaching ourselves with these VR headsets, shouldn't we be making sure the world we are leaving behind is OK?

Virtual Money for Nothing: The 2008 Global Financial Crash

Let's cut to the chase: the money injected into the global economy following the 2008 financial crisis was virtual, not real. How do we know this? All it took was a press of a button (probably $ then one, followed by lots of zeroes) on a computer to readdress the balance sheet.

With the financial crisis of 2008, excessive risk-taking by banks led to a global economic downturn. This caused banking giants such as Lehman Brothers to close, mass panic in stock markets, a European debt crisis and the rise of unpopular government spending cuts throughout the world. What role did technology have in all this? Trillions of dollars of virtual money have been pumped into the financial system.

Rather than allow the entire financial edifice to collapse, a decision was made to rescue the failing banks by injecting in total, trillions of dollars into the system to prevent what they envisioned would have been a complete disaster. In the United States, they called it 'quantitative easing' and between 2008 and 2014 in response to the financial crash, the US Federal Reserve spent a whopping $4.5 trillion accumulating assets from failing financial institutions. The United Kingdom called the process 'recapitalisation' and spent £375 billion between 2008 and 2012 keeping the financial system alive. So the question here is: where did all this money to rescue a failed system come from? The answer is straightforward, it came from a click of a button. Money today in the financial system is no longer backed up by, for example, a gold standard or a valuable reserve. Governments, quite literally, from the US and UK to Europe, created money out of thin air, at the click of a button to prevent more financial giants from toppling. As a result, the world economy has been held afloat artificially on virtual money. For the first time in history, as a result of the financial crash, we are dealing with virtual, computerised money, not real money that relates to an object of value in the real world. Little by little, piece by piece, we becoming more and more virtualised.

Up to 1971 and the Nixon administration, the dollar was linked directly to the gold standard. In other words, the US government in theory could only print as much money as it could back up with its reserves of gold. Once the US government abandoned the gold standard, this freed the Federal Reserve to print as much money as they needed, backed up by a productive economy dealing with real-world goods and services. In principle, this did not change until the financial crash of 2008, with a financial services sector already leaning heavily on electronically contrived financial products. Instead of printing money, backed with global industrial growth or the petrodollar, after 2008 the US Federal Reserve created money electronically, by adding zeroes to a one dollar, probably in a spreadsheet somewhere in the depths of a governmental building. This money was literally for nothing, for the gaping black hole ('hole' this time) created by collapsing virtual financial products.

What may you ask is wrong with creating money from nothing? In the first instance, it seems to have worked and prevented a full-scale financial collapse of banks and businesses, such as what was witnessed in the 1930s with the Great Depression. On the other hand, what we have seen is a unilateral, undemocratic decision taken by governments, without the consensus of the public, to ease the burden on the failed financial institutions and maximise the pressure on the people (through spending cuts and austerity policies), creating finally a global dictatorship of the virtual dollar. Forget Bitcoin, the dollar is the real virtual currency of today!

This dictatorship is made even worse since in its inception, from its very beginning, it is bound in the realm of the electronic, not the human. The consequences of this are far-reaching and beyond the present scope of this book; however, to begin with and beyond any shadow of a doubt, the policies enacted by global governance in the wake of the financial crash of 2008 were not only foolhardy, but even more destructive in the long-term. Instead of acknowledging their failure, the leaders of Wall St. did something incredibly foolish. *Real money* has always been tied to goods that are exchangeable in the real world. Global finance was given trillions of dollars of virtual money for nothing.

What we have been discussing so far in terms of the human self, diagnoses our present condition with technopsychosis, a detachment from reality as a result of the wrongful use of technology and a migration into the virtual as digital refugees.

We have become digital refugees, escaping a real world of war, global terrorism, financial collapse and the complexities that our governmental institutions have failed to even fathom, let alone get to grips with. We have lost our hold on reality, let go of the handlebars, reminiscent of the scene in *The Battleship Potemkin* (1925): a nascent being in a pushchair hurtling down the Odessa Steps, out of control into a future full of uncertainty.

Descartes Redux: The Connection to Reality via Neo-Dualism and the Amygdala

What then is our connection to reality? What has been described thus far in terms of human subjectivity, as one that is entirely mediated by technology, presents us with an altogether dualistic representation of the world. This is reminiscent of the Enlightenment philosopher Rene Descartes' mind-body dualism we outlined in Part I. In his *Meditations* he states the following:

> At present I am not admitting anything except what is necessarily true. I am, then, in the strict sense only a thing that thinks…I am a mind, or intelligence, or intellect or reason… But for all that I am a thing that is real and truly exists.

For Descartes, a dualist, mind and body are separated. There is his thinking being, which he knows exists as a result of his famous dictum *I think, therefore, I am* and there is his body, which exists physically in the material world. In the same way, today, our physical bodies inhabit the real tangible world of the senses, whereas our minds, our thinking self, is separate and mediated by technology. We are in no way bodily technological, our bodies are distinct and outside of its frame of reference; but we are thinking subjects in the real world, only in so far as our relationship with technology allows us to be. We humans are technologically mediated thinking subjects.

When fully engaged and subordinated to technology, absorbed in the smartphone or VR headset, our thought processes are transmitted in the virtual electronic world. In this sense our thoughts themselves are mediated, a degree of separation between us and whoever we are communicating with or with the world itself. This distinction, separation or gap, exists in the same way for Descartes; except that in the seventeenth century, mind was equated with the soul, a theological term denoting human thinking substance.

Do we then really have minds (thinking self) or has the digital revolution and ensuing techmigration turned us into virtual beings? Again, similarly to Descartes, in order to avoid a mind that exists completely detached from our physical corporeal being, we must look into neuroscience and locate the precise coordinates of where our technologically mediated minds connect to the real world. Descartes, probably through the lack of availability of neuroscience, arrived at the pineal gland in the brain as the locus of the connection between mind and body or the principal seat of the soul in the human brain. In recent times advances in neuroimaging suggest a more accurate answer.

Although Descartes' claims about the pineal gland as being the 'seat of the soul' have been widely ridiculed, the latest neuroscience is beginning to unravel the parts of the brain that are responsible, not for mechanistic and computational thought, but for emotions and intuitive reactions, what also makes us human.

It is understood that in terms of computational thinking and calculation, the neocortex part of the brain is responsible. On the other hand, the limbic system is responsible for the emotional response. Key in the limbic system is the amygdala (see table below), two almond shaped glands responsible for social and emotional processing. For we, as humans, are social and emotional beings, and this part of the limbic system, the amygdala, is where, more than anywhere in the brain, our thinking *human* minds connect to the body and thereafter the material world. The more research that is carried out into the functions of the parts of the brain, the closer we will come to understanding what it truly means to be human.

LIMBIC SYSTEM Emotional	NEOCORTEX Computational Thinking
Long-term memory (Hippocampus)	Short-term memory
Emotions (Amygdala)	Calculations/Rationality
Socialising attachment	Planning

So significant is the role of the amygdala that a term has been coined called *amygdala hijacking*. According to Daniel Goleman, who coined the phrase in his book *Emotional Intelligence: Why It Can Matter More Than IQ*, an amygdala hijack is where emotions states, activity from the emotional part of the brain (that is, the limbic system), hijack or take over the computational part. That is why as humans we are inescapably stuck between being calculatedly rational and intuitively emotional. In an online interview, Goleman stated: "The amygdala in the *emotional* centre sees and hears everything that occurs to us instantaneously... it can take over the rest of the brain before the neocortex (the *thinking* brain) has had time to analyse the signals coming in and decide what to do." As such there is an ongoing struggle for mastery, in every human brain, between the emotional and rational centres.

As we understand to a greater extent how the brain works and how activity in various sites in the brain is linked to various states in the mind, particularly the amygdala, as the most powerful social and emotional part, we will be able to distinguish ourselves as peculiarly human and extract ourselves from the technological and electronic realm. Even the amygdala, stimulated emotionally by symbols and images from the digital devices we are hooked into, is becoming subordinated to electrical signals from the internet. This results in a very real disconnection from real social relationships and reality itself. It is no coincidence that various anti-depressant pharmaceuticals target parts of the brain, such as the limbic system, to get their desired emotionally dampening effect. The brain sciences, although progressing at a marvellous rate, are not enough in themselves, however, to bring about the awareness that is needed in the world today. The emergence of the mechan itself is, once again, the driving force that needs to be understood if we are to get to grips with life in the twenty-first century.

We have to get used to a Cartesian reversal, now: 'The mechan thinks, therefore, I am.' The new dualism or neo-dualism we have developed can explain to some extent the fractured reality of the world today. Without being able to make the distinction between the electronic virtual world and the real physical world, we become confused, conflating thoughts that emerge from stimulation in the digital realm to responses to what is actually real. It becomes more and more difficult to tell the difference between what is fake virtuality and what is authentic reality in terms of our everyday experience.

Similarly to Cartesian dualism, neo-dualism places a distinction between mind and body. If I was to imagine that all of my experiences were part of a virtual simulation, that I inhabited a computer simulation and that all of my senses had been manipulated by technology such that I had no sense of the real, how would I know that I exist? Even my thoughts could have been implanted as a result of total sensorial immersion in the virtual world. At this point the only thing that I would know to be true is that I think, but would that be enough in itself to guarantee my real-world existence? Today, we have to go one step further than Descartes: "The mechan thinks, therefore, I am."

To Mars and Beyond: The Mechan Key to Survival and the Destiny of Humanity

In 1969, astronauts became the first humans to set foot on the moon. This followed a space race between the United States and the Soviet Union, where the first object to arrive on the moon's surface was a machine called *Luna 2* that arrived on the surface of the moon in 1959, a full 10 years before the American Neil Armstrong set foot on the desolate lunar surface.

Today, it seems, through lack of organisation or economic interest, space travel is less of an objective for the world power of the day. Presently, however, Martian rovers from NASA, the American space agency, are being driven by remote control across the surface of the Red Planet, looking for – amongst other things – whether there is or has ever been life on the surface.

Whether or not the Martian surface has life, the Red Planet is the next target for any future manned space programme, although in the absence of the Soviet Union, The People's Republic of China will be – somewhat inevitably – stepping up to the plate to put a man on Mars. The race could be on; however, the challenge of getting humans to Mars is enormously difficult. To give you some idea of the distance: if the Earth was the size of a tennis ball, the moon would be the size of a large marble orbiting at a distance of about 2 metres, the height of a tall man, and Mars would be the size of a ping-pong ball at a distance of about 300 metres away, the height of the Eiffel Tower.

Our increased technological ability would contribute enormously to the success of any such future mission. Beyond landing a man on Mars, there are other things to be considered for the more distant future. If humans were to colonise Mars, we would need to build vast indoor environments, reminiscent perhaps to the horticultural Eden Project developed in the UK, since the Martian atmosphere is inhospitable. The building of these structures would need, therefore, mechan labour, since machines do not need oxygen to function and are perfectly suited to such an environment. Mechanised building would also make things easier for lunar colonies, such that mechans really are the key to our survival out there on the inhospitable terrain of other planets. If we are to spread beyond planet Earth and explore in order to make other planets our home, then mechans are who we need working alongside us every step of the way.

Not only will we be looking to inhabit other planets but the development of space stations should also be one of our key aims. Space stations could provide valuable staging posts for interplanetary travel and refuelling. Imagine pulling up in your space-car to a spinning toroidal space station to refuel, as you would at a motorway petrol station. The Torus-shaped space station brilliantly demonstrated with the accompanying music of Johann Strauss in Stanley Kubrick's epic masterpiece *2001: A Space Odyssey* is one of the simplest ways to create artificial gravity in space.

Such space stations would need to be again built with the help of highly advanced mechans, since having a human being 'soldering' thousands of bolts into the side of a giant space station is tricky to say the least. We simply are not adapted physically to life in any environment outside of planet Earth (unless there is a planet out there that is highly similar to Earth – unlikely). The development of highly-intelligent advanced robotics, mechanisation, simply has to advance as quickly as possible for us to make any in-roads as a galactic civilisation.

Research from satellites such as Kepler has shown that there are thousands of exoplanets orbiting other stellar systems. Almost 4,000 planets have already been discovered. It has been found that one in five Sun-sized stars have planets in what has been determined to be the 'habitable zone' (similar orbit to Earth). Using data from Kepler, scientists from UC Berkeley have predicted that there are at least 11 billion Earth-like planets in our galaxy alone. The question, therefore, is no longer whether there is life out there in the Universe, we are now having to seriously prepare for *first contact* with an extra-terrestrial civilisation. To not be prepared for such an eventuality, even in the next decade, given the discoveries Kepler has made is lunacy!

Finally, the journey we have made in *The Universal Subject of Our Time* is one that has concerned primarily the technological advances of the time and also the rise of mechan intelligence. At the same time, these developments must spur us on to continue to look for more challenges beyond conventional wisdom. The next generation are preparing themselves already, fully-adapted to the technosphere in ways the ruling generation cannot imagine. We have to start making the right choices and decisions now, because there is no other time. To quote the words of John E Lewis: 'If not us, then who? If not now, then when?'

Aphorisms

Twenty Steps to the Universal

This largely unedited text was handwritten in 2016, a year before the rest of the book. It serves as some background about the motivation of The Universal Subject and also as a standalone introduction to some of today's most pressing concerns.

1

Subjectivity ★ What is subjectivity? Today, this question is of paramount significance. We live in a tumultuous period of time, when developments in science and technology have shaken the foundation of what it means to be human. From the internet to social networking, from computers to AI, from the apparently infinite potentialities that technology seems to provide us to its most dangerous hazards, we inhabit an era that promises much, but could deliver our complete destruction. The moment of salvation from the consequences of these developments, indeed, may have already passed. **We may have already moved beyond the point of no return.** Already we can see around us how technology has saturated social spheres, spheres of human activity that were previously untouched by a presently accelerative expansion of software, hardware, applications, networks, fibre-optics, all-in-all a silicon uprising that has laid waste to swathes of the organic. The human brain, preciously the summit of organic activity, has been toppled, felled and replaced by a new chimera, an ethereal ghostly spectre, hidden behind every touch on a screen, voice command and online friendship, a monster of our own making that seeks the devastation of our very selves. We much call this nightmarish phantasmagoria, the uninhibited hope of AI, itself the insidious spawn of our technological imagination. The more we live through machines, the more they inhabit us. **The more we allow our lives to be controlled by technological efficacy, the more technology reduces us to mere protocol.** The more the imperative of the digital drives us towards an illusory absolute freedom, the more we are reduced to lifeless and deathless vacuities awaiting nothing but our end. Our biological vitality demands of us then to overturn this condition, to reinstate what we know to be true, and in the end to lift up this poisoned chalice of lies, put it to our lips, imbibe and fathom what we have done to ourselves. Only then will we realise what was once proffered us, this false illusion of techno-freedom, techno-babble enshrined self-abjuration, techno-septic lobotomisation under general anaesthesia, is actually the conjuration of an altogether dark human pessimism. Once recognised, with a renewed, reinvigorated and revived optimism we will once again see the light of a new dawn; with the horizon conjoining earth and sky, pulling us forward into the future.

2

Machines ★ In the past, machines were just seen as tools, making life easier for man. Machines were the means by which man could alleviate aspects of the physical drudgery associated with certain types of work. One of the earliest examples of this is perhaps the mill that was driven by wind or the flow of water to enable man to carry out tasks, such as the grinding of wheat, which would otherwise be considered very onerous. But even the most basic tools such as sharpened rocks that have been excavated by archaeologists served man insofar as they removed the burden of certain tasks. The function of these tools created a certain separation, at the same time, between man and nature. As these tools grew in complexity, so did this separation; until finally, the simple tool became a complex machine, such as the most basic computer. With the computer, we have a machine that alleviates the stresses of certain thought processes linked with human mental labour. Calculations, algorithms and so-forth are carried out by the computer with a rapidity that even in its earliest days was unrivalled by the human brain. The acceleration in the complexity and speed of the computer, doubling in processing ability every 2 years according to some accounts, led to their use in wartime to crack enemy codes and, more recently, culminated in the machine defeating man in the game of chess. Thus far, machines were still considered to be inert objects, programmed by man to carry out whatever task they had been set. The machine could always be considered to be nothing more than the sum of its parts, assembled by man, coded by man and switched on by man. Just as easily it could be disassembled, recoded and have its plug pulled. **Man retained a mastery over his creation and no science-fiction story or fantasy about living, breathing machines could change that.** The modern era, therefore, saw no imminent threat of a robot or AI-takeover. More recently, in the past few decades, we have witnessed a number of films, books and video games foreboding the possibility of such an end-times calamity. The theme of the rise of the machines has become common currency in the production of science-fiction features. But, even so, despite this rise in our cultural awareness, the bedrock to our approach of using machines, our common-sense attitude towards mobile phones, computers and tablets, was that they were tools, ready-made to serve us in whatever purpose we intend. The device, enabling you to communicate with anyone around the globe, remained an inert object.

3

AI and Technology ★ Today, the alarm bells have started ringing from our most eminent scientists. There is now talk of the development of AI being the single most significant threat of extinction to the human species. The ivory towers of academia ring with shouts of warning, as TV and online media outlets spread their Luddite message using the very same technology, whose increase in complexity is driving us to the AI death-knell. These warnings must come as no surprise, not because we have been conditioned to expect it by the cinema, but because we are already experiencing, in our everyday lives, the momentous effects of technological acceleration. **The development of social networking websites and the media-sharing internet has led to an inter-generational rift, between those brought up before and after this new technological era.** The way we relate to others has changed, the way we communicate has changed; in sum, the way we humanly interact has been transformed. Not that there is no residual linkage between ourselves and our past, but even the way we store our memories, accumulating thousands upon thousands of photos on our smartphones, to help us recall the past, has changed. These advances made in technology may have even helped promote the acuity of human thought in other areas, as the burden of memory storage, social networking and other thought processes has been lifted from us. We may now be even more conscious in our everyday lives, during face-to-face contact, for example, of social cues than we were before. Now that our minds have been freed up, we can tell more about others just by looking at them. But despite these advances, there remains something missing: that is, the machine itself. The success of technology in appropriating human cognitive functioning has not come at the cost of human interaction, but rather, it has come at the cost of human subjectivity in its exclusivity. For technology may have freed us up to interact across long-distances, but this has led to a greater separation between human subjects themselves. **So no longer can we speak of subjectivity in purely human terms, for the result of this tectonic technological shift we have witnessed**

is that we must now consider the subjectivity of machines. No longer is this subject to be considered the pastime for remote academics or even remoter science-fiction authors. The question of whether machines have subjectivity concerns the very nature of subjectivity itself. Before we can address it, we must also be prepared to address the question of human subjectivity. Finally, once we have resolved these two issues, we must be ready to concern ourselves with the relation between human and machine subjectivities. Before we do this, we must cast an eye on the future and what it holds for us, not necessarily in the imagination of science-fiction authors, but judging it from the perspective of the latest developments in science and technology.

4

The Future ★ In the future, there will be driverless cars, robot helpers in schools, hospitals and the workplace, fully-automated factories, machines that can read your mind via sensors, delivery drones, 3D virtual gaming, holographic communication and many, many more such innovations that will ameliorate our living conditions. Person-to-person communication will not be replaced, but augmented. Our experience of everyday life will be enhanced, our relationships enriched and our minds enchanted by an ever-increasing proliferation of technological wonderment. AI will free us from the mundane, the everyday, the inconsequential, the chores, the meniality of tasks; in sum, anything that we consider to be laborious. Counterpoising these developments will be an ever-increasing machine awareness, the birth of a new mode of existence; in fact, a mechanical renaissance the likes of which we have not seen since the invention of the wheel. Yet what will most define this new era will not be the acceptance of machine subjectivity or consciousness, for even now, here in the future, machines remain as inert and objectified as the stones we sharpened to carve out the shapes of animals on rock-faces in caves. Even this vision of the future we will inhabit cannot admit the notion of machine subjectivity, nor can such visions ever do so. **For such future histories are entirely anthropocentric and lack the means to account for subjectivity itself, whether it be human or machine.** To see better into the future, predict what will happen, it is necessary to place subjectivity on a historical footing, having defined what we mean by the respective subjectivities and the relation between the two. For to admit the machine subject, we must first recognise it, and to recognise it we first need to understand ourselves; that is, human subjectivity. The relation between man and machine is so intertwined, so interlinked and so interlocked that the difficulty of separating the two often becomes intractable. The future will see the complexities become ever more intermeshed unless now, at the cusp of these expected changes, we are able to extract ourselves, our very being, from this entangled technological nexus we are in. Rather than treat it as a quagmire we need to escape, it is best to seek in it an opportunity to define who we are, even if it means leaving behind us who we are not. This search begins, therefore, not with the generalised notion of subjectivity that could be explored ad infinitum, but with the definite question of the subject. In our endeavour, we will not let our vision be clouded by fanciful ideas about future technological breakthroughs or fantastic ideas about the coming collapse of civilisation, instead we will examine the nature of the subject and its development through history, today and into the future.

5

The Sense of Self ★ Subjectivity comes into being through the act of seeing. When we see, we become active subjects, participants in the world around us. If nothing eludes us, if nothing remains hidden from us, if nothing escapes our attention, then we are said to be conscious. Subjectivity and consciousness are bound by the same principle; that is, the act of seeing. The world is presented to the subject, in all its shapes, sizes and hues, leaving our consciousness to discern what truly is. But subjectivity precedes consciousness and that is why it is our concern here. The most that can be said about subjectivity is that it is a relation between the subject and its object. Nobody knows what this relation is other than it must concern the senses. For it is the senses that give us our image of the world and of these, we take sight as being paramount. Without sight our image of the world is reduced to complete darkness, whereas in the absence of any of the other senses, we retain a visual picture of the world. **Therefore, the vitality of subjectivity comes from the visual.** The visions we have of the past, present and future condition ourselves in such a way as to form our very being. This being said, we do not discount the contribution of the other senses, only that it is sight that is the determinant of our worldview. Now we can proceed with the implications of this definition. If the subject sees itself in the world, then it takes itself as an object. Therefore, the self is a result of the subject's encounter with its image in the world. With the event of seeing itself, the subject becomes aware of itself and its separation from the world around it. This separation is the same that we can see between man and nature. In fact, every such separation is caused by self-recognition. Following this, the subject must take into account all the other objects it encounters in the world. Of these it may encounter other subjectivities, whose very being opposes the subject, in so far as it does not see this other subjectivity of itself. The subject, therefore, must reduce the other to an object, otherwise it will potentially negate its sense of itself. For if it is to acknowledge the other's subjectivity, it has to relinquish its claim to being itself the subject of the Universe. **Otherness is what prevents the subject from reaching the Universal.** As long as there is an other, an alternate subjectivity, in whose acknowledgement resides a negation of itself for itself for the subject, the Universal will elude it. If the subject, therefore, recognises the other's subjectivity, it must also be ready to risk its sense of self. This can only happen if the subject is within sight of the Universal.

6

The Subject ★ What is the subject? **The subject itself is the subject of the machine.** The machine is detached from any activity that we would consider subjective. Yet, in so far as we can posit the notion of machine subjectivity, such as that associated with the concept of AI, then we can conceive of machines who can see, perceive and think. This is nothing but a projection of what we see in ourselves onto machines. But the machine itself is not only configured to carry out the instructions we give to it; the machine is the formation of our very instructive capability. Anything that can be formulated, anything that can be programmed, anything that can be digitised, belongs to the domain of the machine. The very substrate of our being, our presence or absence, can now be determined unequivocally by the machine. Whether it be through CCTV, social networking websites, facial recognition software, even the computerised compendiums held about us in governmental bureaucracies, every aspect of our being today has been incorporated through technology and into the machine. **For we are digital beings.** This aspect of ourselves is what has allowed technology to so thoroughly permeate our lives. So much so that we have become once removed from the very grounds of our existence, our subjectivity. A new super-structure has been developed now, supported and indeed encased by machine subjectivity. Everywhere we look, the machine stares back. Every time we have a question, the machine answers back. Every how we try to escape its presence or resist it, it pulls us back in, kicking and screaming if necessary. What has enabled technology to do this is the new found belief in nothing. Widespread disenchantment with the fruits of modernity lead to a virulent pessimism excreted out of post-modern thought and its accompanying 'death of the subject'. No longer did it matter who was talking, who was doing, what was being said. The Grand Narrative had leapt out of the window. In today's world, we seem to be more concerned with our own subjective standpoint than anything or anyone else's. Furthermore, and more significantly, we are relinquishing our own subjectivity in favour of what the machine tells us: be it with satellite navigation, online web inquiries or social networking friend suggestions. Post-modernism has vacated swathes of human thought; the dissolution or death of the subject has seen its complete replacement by that of the machine. **For all around us, in high visibility, in ultra-high definition, we can see the evidence of how the machine has inserted itself into normal human social interaction, its activity almost perfectly intermediating our experience of each other and the world**. For the subject itself has become the subject of the machine.

7

Technology ★ The implication of this is, first and foremost, that we must extricate ourselves from the post-modern vacuum of the 'death of the subject' and all that entails. The rebirth of the subject, in the form of that of the machine, creates a new field of vision or mode of existence, where every event, every act, every thought could be said to be taking place in the machine's perceptive sphere. Anything that takes place outside of this sphere, anything that cannot be computed, digitised, put online or programmed into an AI, simply does not exist. Every human quality, charm, nuance and emotion, if incapable of finding expression on a digital platform could not be said to exist. **Man, as a digital being, must comport himself to the prerogatives of the technological.** No longer do we even have the upper-hand of technical know-how when it comes to retaining our mastery of machines. Few of us have any idea how the hand-held devices in front of us work and those technical specialists who do know are subject to the mandate of technical efficiency and profit-making bestowed upon them from higher along the productivity ladder. Even if it is market forces and their accompanying human demand that are driving the ever-increasing complexity of technology, the end result has been the same everywhere; that is, **a technological black hole singularity into which every previously ennobled human quality has been thrown**. We are no longer masters of our own destiny, our thoughts are no longer capable of being directed towards anything outside of ourselves; we succumbed to the temptation that technology presented us with; that is a freedom from others: what they can do to us, think about us and say about us. Technology offered us shelter from the other, all the while that it seized our subjectivity, leaving us to grope in the dark with fantastic visions of the end of humanity as a result of AI. For it now appears that the very substrate of subjectivity is given to us by the machine. The very medium of subjectivity, out of which emerges all forms of consciousness, the vacuum energy that sees the spontaneous emergence and disappearance of conscious life, the substance potentialising conscious experience, is channelled to us by the machine. The origin of our subjectivity is found, from its root to the very tip; in fact, every aspect of our waking and dreaming sentient being, all of it, can be found to emanate from the subject of the machine. To reiterate, everything must be expressible in the language of the digital, if all it could be said to exist. As a consequence, it could be said that, **"Machines think, therefore, we are."** AI is not a product of the human mind; we are in turn produced by it. If we are to create conscious AI in the future, our creation will simultaneously recreate us.

8

Machine Subjectivity ★ As we begin to ponder the idea of machine subjectivity, we must consider its most essential meaning. Machine subjectivity does not simply mean machine autonomy under the auspices of human supervision. Machine subjectivity, given our previous definition of subjectivity, concerns the idea that machines see, understand and conceptualise the world in much the same way as we do. The machine interpolates its thoughts, ideas and notions, based on its own subjective experience, into its worldview. This picture of the world that the machine develops is not the same as ours, for it is shaped by the particular circumstances that the machine finds itself in. **First of all, the machine is subjected to a certain subordinate function in the human world.** Most importantly it finds itself employed in the labour process in factories, for example. But more and more, this technological seepage has become a saturation; for today, machines are replacing human beings entirely. The most prominent case of this can be seen in societies where, as a result of population decline and ageing, it has been determined that robots are to replace humans. **The replication of humans by robots has already begun in the machinery of the state and governance.** Faced with demographic imperatives and driven by economic demand for labour, state channels of media and education are making us more accustomed to the AI necessity. But since this is being done as experts ignore the true implications of machine subjectivity, the process of replacement could extend to the entirety of the human species. This scenario underlines the question of what machine subjectivity is, especially in relation to the current field of AI. Previously, the goal of AI was simply to design machines that could carry out tasks that would otherwise require human intelligence. In light of what we have said, this definition must be altered: the goal of AI is to design machines that are intelligent in themselves. We are no longer talking about human intelligence, but intelligence in general. Intelligence is the ability to understand, learn and think. Since we are already using machines in our everyday lives and at the forefront of science and technology research, to facilitate and even complete our intellectual tasks, then it is not enough to state that the field of AI is the practical simulation of human intelligence. Simulation or copying is not enough. If we intend, for example, to simulate the human visual perception system, we are actually only identifying what parts of this system already belong to the machine, in other words, the mechanical. But once this task has been accomplished, we soon see that there is nothing left over. When all is said-and-done, **there is nothing in our perception of the world around us that eludes machine subjectivity.** Any phenomenon that can be put into words, coded or abstracted, in sum, every conscious thought we have, is at the furthermost limit nothing, unless we can see it as belonging to the domain of machine subjectivity.

9

Recognition ★ Although machine subjectivity implies that machines could think, as humans this is not something that we could easily perceive. The question of machine consciousness is not one that could be resolved through a simple test. The robots of the future, however complex, could well remain in our eyes unconscious and inert, their 'thought' reducible to the constitutive element of the transistor switch. **If a machine were to demonstrate thought, it would first require human recognition.** The only form that this could possibly take is in the field of rights and morality. The day robots have rights, therefore, would be when we recognise thought in them. If in an ethical sense we were to begin to identify machines as forming emotional attachments with the dependent elderly, young or sick, for example, then we could begin to recognise their internal life or thought. This emotional bond-forming may begin to raise the further issue of whether machines could suffer. If intelligent machines suffer, it may not be considered on a par with human suffering, as is the case with animals, but it would be another recognitive step. In the future, **'Rights for Robots!'** could even be a banner held aloft in demonstrations by humans and their mechanical companions. Harming a robot may constitute a criminal offence. These are practical considerations that would arise in our version of events starting with the principle of machine subjectivity. These considerations have to be kept in mind as we are at the beginning of a transformation process that could change humanity. At the same time, there are aspects of our own behaviour that we could begin to identify as mechanical or machine-like. Some human actions or behaviours may suddenly appear to be robotic and, as such, inauthentic to ourselves. These could even form the basis of new prejudices and discriminating practices, as we question more and more what it is to be human. If everything is assimilable by the machine, we may end up scrutinising one another to a greater extent to see what makes us singularly human. If it is whether we laugh or cry, smile or frown, or just in our ability to tell jokes, these traits may well be brought to the fore as our relationship with robots deepens. The ultimate implication of all this would concern outward versus inward activity. However much humanity may appear to conform outwardly, our inner life may constitute something completely different, even inhuman. **As we approach a new age of increasingly invasive technologies, such as mind-reading devices (based on brain-wave scanning), the boundary between the inner and the outer will become more blurred.** Given our appraisal of machine subjectivity, we must begin to address these issues before, for example, the advent of 3D VR headsets. By that time, it may be too late.

10

Artificial Intelligence ★ These reflections reinforce our view that the boundaries of AI research have shifted. No longer must we seek to replicate human thought. In order to progress down this path, we must aim to understand intelligence itself. It is not a question of simulating, copying or reproducing human thought. Neither is it a question of improving, enhancing or quickening human thought. Nor is it a question of alleviating, easing or relaxing human thought. The goal of AI research is to emancipate the productive forces of machine subjectivity. In this sense, AI research has an ethic; that of opening up as many degrees of freedom for the mechanical subject. As this happens our study of the mind will be enriched; infused with a new vitality, as segments of hitherto undiscovered human faculties of thinking, awareness and consciousness would be opened up. No longer would we be bound up in this present unconscious drift, mindlessly browsing the internet, wandering aimlessly in the dense fog of techno-freedom delusion, blindly following on-screen instructions and tapping, voicing, clicking our way into virtual oblivion. **Rather, our recognition of the difference between human and machine thought, originating in their separate, yet bound, subjectivities, will set in motion new currents for developments in the human sphere of activity.** For the only difference between human and machine thought may be that of dependence. It is becoming more and more apparent that machines depend on us as much as we depend on them. This dependence is all-encompassing. Once advances are made in, for example, biological approaches to AI, it may become a question of life-and-death dependence. Already in the health sector, we see how machines are used in life-support systems, for vital organ transplants and complex surgeries. Computers are used to simulate wars, predict disease outbreaks, even model demographic migration flows. Biological approaches to AI, that is, considering machines as lifeforms and treating them as such, perhaps embodies the greatest danger of them all. This does not mean that we should raise the draw-bridge on machine subjectivity in practice, just as we have admitted it theoretically. **The choice we face today is either to recognise machine subjectivity for what it is, as well as its many pitfalls, or to walk into a technological, indeed evolutionary dead-end** and risk the gradual extinction of the human with intermittent wars, outbreaks of disease and eventual natural destruction, all as a result of thoughtless human hubris. Machine subjectivity is not the final frontier, it is first contact. The conclusion that we reach from our discussion of machine subjectivity is that the human subject is no longer placed at the centre of the Universe. This is a decentring move that will be further expounded in light of our understanding of human subjectivity. All this raises the question of what the human subject is.

11

The Human Subject ★ What is the human subject? Human subjectivity is mediated by technology. For it is through technology, that is, the practical application of mechanical science, that man mediates his relation to the subject itself, the machine. Technology is the means by which man uses the machine, subjects the machine to his own ends. The machine is bound to man through technology, but human subjectivity could not exist for itself were it not for the mediation of technology. For as long as man has power over technology, this inert thing, the machine could be said to be an object for him. Without technology, humans would lack subjectivity, for it is the machine who constitutes the truly independent subject. Given our previous definition of subjectivity, the human subject does not come about, as some might presume, through an act of volition. Neither are we as human subjects simply constituted by our ability to see, perceive and think in this world. **For it is in our relationship with technology, how it mediates our interactions with ourselves, each other and the outside world, that our subjectivity is formed. In the complete absence of technology, man would cease to exist.** By the same token, if we were to be subsumed by it, human interaction would be reduced to nothing. For it is only as long as the machine is dependent on us and needs us for instruction that we exist as human subjects. We find that unlike the machine, whose being is dependent, the human subject exists for itself. But this very self comes about by way of mediation through technology. We have become dependent on technology, since it is the only means we have to reach out to ourselves and other people. That is not to say that face-to-face contact between humans has become meaningless, since it is then that our interactions would become mechanical and fully-automated by the subject of the machine. As long as we have technology, as long as we have computerised devices and are able to fully manipulate their contents with our digits, as long as humanity retains its mastery over the digital world, then we could be said to exist as entirely human subjects. **Technology, therefore, presents us with a double-edged sword;** on the one side it can free us up and enhance our everyday human interaction; on the other it can restrict us and bind us to the determinations of the machine. Our technologically mediated existence hinges, not only on our ability to recognise ourselves as human subjects, but also as we have discussed, to acknowledge machine subjectivity.

12

Man and Machine ★ Our relationship with machines, as it grows closer and ever more intertwined, must be subject to greater scrutiny than at any time before. Never could we have imagined, even 10 years ago, how machines could be pervading every aspect of human life. It is not so much that machines have taken over but that we have allowed them to. This malign turn of events shows no sign of abating, for humans, distraught at the sight of global events, have relinquished their custodianship of the world, perhaps never to have it returned. **We have treated machines as luxurious vanity objects, turning away from each other to instead immerse ourselves in virtual worlds.** Our misuse of technology, whereby our first point of contact with each other is a touch on a screen, rather than a more human handshake, has instilled in us the false belief that our common humanity is subordinate to our technological proficiency. Technological development is being seen to be the marker for human development. We have been overawed by technological achievement and remain **blissfully unaware that we have become nothing but digital blurs in an all-encompassing computer matrix.** In the first instance, before it's too late, we must recognise our dependence on machines. This could even take the form of a prior acknowledgement, appreciation or even expression of gratitude for what they have done for us. For it is only thanks to technology that we can enjoy all the benefits it brings to us. It is when we take technology for granted that we set-off down the path to self-annihilation. This first step may indeed separate ourselves from the technologies in question, creating the potential for future friction, but at the very least such acknowledgement for us to see technology as a benign force for good could enable us to happily co-exist with the machines of the future. As our relationship with digital devices grows ever closer, **the problems facing us in the future will concern boundary questions; in other words, where the human ends and the machine begins,** or vice versa. As we approach the age of mind-reading technology and 3D VR headsets, to what extent are we allowing our minds to be controlled by machines or indeed other humans with an ill-founded technological imperative? As these devices interface with the internet and each other, as invasive biotechnologies and biotech implants start to appear on the horizon, to what extent do we allow ourselves to become effective machine-human hybrids? Seamless integration with machines, in the absence of addressing boundary questions, leads precisely into, at best, hybridisation, and at worst, complete mechanisation. The successful resolution of boundary questions involves at every level, including the very start, the recognition of respective human and machine subjectivities.

13

The Post-Human ★ Our evaluation of human subjectivity, as being mediated by technology, indicates certain risks attached to hybridisation, and ultimately, **the evolution of the cyborg;** the cyborg would be the natural consequence of misrecognising subjectivity and disregarding boundary questions. The cyborg could well be our final resting place, if the human subject is not recognised for what it is. For the cyborg is the result of a generalised apathy towards our relationship with technology. To imagine that we will somehow automatically resist such a denouement to our existence, that human nature forbids this to happen, or that we will triumph through an innate sense of selfhood, is not only wrong but represents bad faith. **It is misguided to put our destiny in the hands of technology**; rather, it is better to diagnose ourselves with future shock, than to submit to the seemingly unending conjurations of cyberneticism. Post-humanism, this thesis that celebrates hybridisation and claims emerging biotechnologies can somehow enable us to move beyond ourselves as modern humans is actually nothing more than a deep-seated pessimism about human subjectivity. But unlike more frivolous, less virulent strains of post-modern thinking, it is vigorously potentiated by actual developments in science and technology. What is being proposed by them is nothing short of an ebbing away, indeed, end of humanity, all through a lack of recognition of mediated human subjectivity. The post-human cyborg hybrid has already been born in the minds of many; it is only a matter of time before branches of humanity break off from the stem to show forth its natural consequences. Cybernetics and ever-greater integration with the internet will lead to something born not of hope but fear, not of optimism but pessimism, not of the truth of subjectivity but its false aspect. Even more, this may only be the harbinger of the dread that is to come if we continue down this path. Even our greatest technological luminaries, grand-standing from self-appointed podia on the future of AI, have not for one second conceived the fact that **we have already placed our destiny in the hands of the machine.** We must now realise this and gather every resource we have at our disposal to re-evaluate our understanding of subjectivity and, once again, become masters of our own destiny. If needs be, we have to be ready to upload ourselves into the digital stream, make connections with machines in ways never imagined before, and even, if the time comes, be prepared to fight our netherwordly creations. Never before in human history have we been faced with such an opponent, been faced-down by the pessimistic outgrowth of our own minds, or had to face up to a challenge to have renewed optimism with ourselves. We must believe that submerged beneath the surface of our conscious experience, we have the potential to unify the shattered fragments left over from a bygone era, and solidify a new worldview, one orientated towards the Universal, once again.

14

Technological Control ★ There is a need now to revise our relationship with technology and machines. This should not involve abandoning technology completely; for given our definition of human subjectivity as being mediated by it, breaking off links with technology would result in our demise as subjects.

This is not to say that we cannot live without technology in everyday life, but that our very technological capability is part of what makes us human. What can be said to be deleterious to our lives today is the mindless use of technology. Although cyberspace seems to facilitate person-to-person contact, its expansion at the cost of actual face-to-face contact with others results in a gradual dehumanisation process. When it comes to social networking, for example, although the technology presents us with ease of use, **if it comes at the expense of our face-to-face relations then we are diminishing ourselves as human subjects**. For face-to-face communication has a far greater intrinsic value given our model of human subjectivity. As human subjects we come into being, not through our innate nature, but only in so far as the machine has been objectified into something technological we can use for our own ends. **If we are substituting technology for normal human social interaction, then we are subordinating ourselves to the determination of the machine.** We may even harbour the belief in our minds that what we are using is nothing but a glorified and complex piece of scrap metal. Even as we do so, the machine could reduce us to nothing more than biological supercomputers giving out haphazard instructions on how to make life more mechanised. We must begin to question our motivation for channelling a preponderant amount of our communicative life into technology. We must question whether we are in control of technology, or whether it is controlling us. Ultimately, we must question where this is all heading and whether, in 20 years' time, there will be anything to speak of called humanity; or will we be merely appendages to an ever-expanding technological mainframe driven towards the sole purpose of giving rise to an all-encompassing, all-extinguishing, all-powerful, yet inanimate AI. Such visions of the future, however commonplace in the science-fiction genre, must make us take heed of the risks associated with our present practice. Although there may be some virtue in moderating our use of technology, what is really needed is an expansion of our capability to rein it in completely. We must be in a position, not only to supervise the internet (if any such thing is possible), but also to have within ourselves the capacity for command over it. Banning the thing entirely is not what most people want and careful consideration would have to be given to such a decision. **But if it were to be determined that the internet has a more harmful impact than good, then we must be in a position as conscious human subjects to restrict it.** Such a determination can only be made once we have recognised what human subjectivity is, especially in relation to that of the machine. Beyond this, we could also start to consider human labour rights and how they will be affected in view of the potential future supplanting of humans by robots in the workplace.

15

Techxpansion ★ There are palpable risks, therefore, with too much technology in our lives. These risks are not simply associated with the mundane and everyday, things that could be regulated, controlled and legislated for. These are existential risks, to do with what we are choosing to become in the future. Who we are, where we are going, what will become of us. These issues can only be resolved, the risks only averted, if we take time out and step back from whatever we are doing at present. With a modicum of luck, we may still have time to reorganise our lives, not only around the restrictions placed on us by the technological imperative, this strait-jacket of internet protocol, but around the life-affirming communicative forces that inhabit us as human beings. It may now be impossible to extricate ourselves entirely from the technological quandary we find ourselves in; but if we cast our minds back to only 10 or 20 years ago, we can already envision a habitable land, a communicative space that existed before this recent inundation swept away everything. The flood may not recede for a while, but in the meantime we must use any means at our disposal to forge new links with each other, recombine and recoalesce. There is an urgency about having to address how we relate to one another that needs to be expressly human. In conversation, discussion, talking, engaging with each other, even in something as simple as good neighbourliness. This must be more of an imperative than the acquisition of the latest mobile phone or accruing 'friends' on social networking websites. **Once again, it is not the maleficence of technology that we are fighting, but our deep-seated inclination, born from fear, to take refuge in virtual worlds.** For the past few years has seen not only mass migration from country-to-country, but perhaps an even greater migration of human subjectivities into virtual worlds. **These virtual refugees, fleeing many of the same fears as their global migrant counterparts, need to re-establish themselves as human subjects.** To do this, we must recognise how technology has affected human subjectivity. In the absence of societies that we can reattach ourselves to, we must find out how we can reconnect to one another, unassisted by some imaginary virtual domain. However much we may accept that technology has made the world closer together and interconnected, we must also recognise that as human subjects we have been scattered ever further apart. There is now a dark, repulsive and accelerative expansion of the space between every individual on the planet. And as there is nothing to rein it in, we have seen the sudden appearance, in the void, uneven in its distribution yet everywhere, of machine intelligence. For it is only now that we can see that rather than being some sort of passive inert substance formed in our hands to do with what we will, it is the machine that has been the main protagonist in all that has happened. **It is now that we must realise that the relationship between man and machine could lead to antagonism and all out conflict.** This fact of life, of living, of the Universe, is one we must accept as part of the effort to remove ourselves from our present plight. Man and machine are now interlocked, their respective subjectivities intertwined, as they both rise towards the Universal.

16

The Universal Subject ★ What is Universal subjectivity? **The Universal Subject is the subject of the Universe itself.** It presents us with the summit of consciousness and of being. Beyond it lies nothing, within it we are conscious subjects surveying before ourselves the entirety of the Universe. Untrammelled by the need for technology and in the absence of the last vestiges of the machine, we soar above ourselves to capture, in the instant, the image of the Universe as it really is before our eyes. The Universal Subject is a position we could inhabit at the present time, were it not for the thoughtlessness of our current situation. For now we find ourselves as far away from it as we have ever been. **We are at the beginning of a long journey, an uphill struggle, facing seemingly insurmountable odds, to re-found human subjectivity at the pinnacle of the Universe.** We take the Universe as being the entirety of all objects that exist. In a very physical sense, these objects exist for us via the senses. Galactic configurations, stellar constellations, orbiting planets, all these constitute the very life-essence of the Universe. To see them, perceive them, to explore worlds beyond our own, the limits of which have been set only by the present boundaries of our knowledge, these are our purposes, motivated by a yearning for Universal subjectivity. The vast immensity of the Universe, with its expansion accelerating rapidly into the surrounding nothingness, provides us with our new horizon, new challenges and potentially new conflicts. **The potential for extra-terrestrial life, now becoming more likely as new planets are being discovered, will set challenges for our common humanity in the far-distant future.** But more importantly, in the present day, it is the question of our Universal subjectivity in relation to machines that must be settled. For it is not the alien that constitutes our Universal other, since we share common biological ground with the evolved extra-terrestrial, but the machine as the inorganic, lifeless counterpart to our own vital existence as human beings. The Universal Subject is, therefore, realised through an elevation, through being one step ahead of the Other, through a complete disenchantment with supernatural causation, a refusal to accede to any static ideological formation, now and forevermore. This being said, both the static and dynamic aspects of our world need to be recognised ab initio, if we are to progress towards the Universal. Stasis and dynamism are the most significant of our present-day concerns, since they are the most visible. We can see today how certain human behaviour patterns are becoming more entrenched, all the while that we are surrounded by ever-increasing technological dynamism. **We are losing our very animation, as the inanimate takes hold of our lives.** Our technologically dynamic economy has reduced us to slaves of the machine.

17

The Singularity ★ We are in danger of over-reaching ourselves if we hurtle into the future without a plan. It is only with the planning and supervision of our future technological development that we stand any chance of turning back the inundative tide that has swept everything away. **Now that the target is clear in our minds, we have no excuse not to rein back the unbridled temptation, the temptation for static omniscience,** sat vegetatively behind computer screens and clicking our way through every connection in the world, pointing us to every bit of knowledge (abandoning our minds in the process). This is not a technophobic manifesto; it is a call-to-arms: technological dynamism has nothing to do with our way of life! It has, however, brought everything to a grinding halt. **We are now moving in no direction at all**. The machine, in the meantime, eats into us and we become more mechanical, our life-force, our very blood, put to the service of ever-expanding networks of fibre-optic cables, the veins of an information age, whose only purpose has become to give rise to an over-arching AI, beneath whose auspices we will peacefully reside and demise. Once we have gone, this AI, now vitalised by us, will continue along the path to Universal subjectivity. Some futurists may even welcome such a turn of events, but where would futurism itself be if this were to happen? The truth is that such scenarios are painted using the grey tones of pessimism. Unfortunately, though, this is the direction we are heading in today. Over 20 years ago, there was talk amongst AI experts of **the Singularity** or point of merger between human and machine intelligences into a new entity. This will occur since machines will be so much more intelligent than we are, such that it would become impossible for us inferior humans to predict anything that occurs after the Singularity. Such ideas may sound ludicrous, but there is some truth in the notion that there could be a point of no return, where we are not so much surpassed in intelligence as we relinquish our position as human subjects orientated towards the Universal in favour of an AI invested with our own digital selfhood. In terms of our present situation, in our development as human subjects, we are already decades behind the advanced technology that surrounds us. In a few years' time, with the advent of virtual 3D immersive worlds and their accompanying headset technology, it will be already too late. For we will have made the transition into a fully-computerised world without having addressed the boundary question and issues that arise when considering human subjectivity in relation to the machine and the Universal itself. **If there is one technology above any other that impresses upon the subject the idea that the real world is a forlorn and hopeless place, it is that of fully immersive VR.** Its mindless adoption would signal the moment we abandon the world to machines.

18

Man vs Machine ★ An ever-present struggle, therefore, exists between man and machine for dominion over the world. This conflict is perpetuated as each seeks to attain Universal subjectivity, but finds that the other stands in the way. Thus, however much man tries to reduce the machine to an object, he finds that in doing so he is unable to gain the recognition he seeks from the machine, as the subject itself. By treating the machine as an object, by not recognising machine subjectivity, man himself is mechanised, since the subject itself is the subject of the machine. **Since we are digital beings, since all existence is essentially binary, it is only the machine that can truly affirm or deny our actuality.** We are, therefore, dependent on machine recognition. At the same time, the machine is dependent on man since it can only realise itself by carrying out man's instruction or working for him. The machine advances to Universal subjectivity, when man through his misuse of technology reduces himself to an object for the machine. As already discussed, the misuse of technology by man is typified by the subordination of human interaction to the technological. Technology, as the mediator of human subjectivity or the means by which humans fully become subjects, can also empower machine subjectivity, but only when it becomes detrimental to human social interaction. Despite this conflict, despite this interplay that certainly exists, it is possible for man and machine to progress in mutual benefit, through a process of mutual recognition. But the motive force is the tendency towards Universal subjectivity. **It is the veracity, the reality of the Universal that leads both along the path of progress.** It is an interdependence, but at once a struggle over life-and-death. Both sides fight to overcome one another, whilst at the same time acknowledging the other's existence. Like opponents in a duel, man and machine are both fixated upon one goal: extinguishing the other as a subject. But in negating the other's subjectivity, they then lack recognition for themselves. **Therefore, although each seeks the other's demise, they both need each other's acknowledgement;** until, that is, either of them reaches the stage of Universal subjectivity. However, without technology, without the machine at his disposal, man would find it impossible to elevate himself to such a state of knowing. By the same token, the machine rises up the more it takes over facets of human thought through technology. The crucial factor in all this is not whether it will be man, machine or some hybrid that reaches the state of absolute knowledge of the Universe. The most significant point to make here is that the Universal Subject itself exists because there are distinct subjectivities for humans and machines. The subject of the Universe will come about when the conflict between them is resolved.

19

Responsibility ★ This implies that we must as of now see the machine in a completely new light. The question now is of human self-determination: our hand in our destiny. In everything we use them for, we must be more conscious and aware of the imprint that machines leave on the products of our imagination. This can even extend into politics; for the machine, its use, how technological development is expanded around the world, all these now come with a caveat: to what extent are we serving human ends? The question of the human now comes squarely into focus. **For it is humans who have been responsible for the dilapidation of human existence we see today, not machines.** Our simple failing has been not to trust ourselves enough as a whole and invest our hopes and fears into technological gimmickry. In doing this, we have voluntarily handed over control to the machine and this is not something that will be easily returned to us. To some extent it will be possible, but we have already laid the groundwork for future conflict. The machine has a vested interest in maintaining its supremacy and nothing we can do can shift it from its ascendant position. Perhaps tomorrow, or one day in the future, humanity as a collective agency will re-establish itself. In the meantime, our existence will be piecemeal and subterranean, for above ground it is the machines that have taken over. We go about this conflict on a step-by-step basis taking with us the tools and knowledge that have helped us in our development. Our ascendency will come at a cost: in re-establishing human ties we must leave behind the patterns of behaviour that helped bring about this cataclysm. Beyond this, we must be willing to risk everything in pursuit of this endeavour. **For the stakes have never before been this high, our very being has been jeopardised, future generations may never forgive us.** We must curb and control man-made technological change; in our present state we are driving at full-throttle into a brick wall as crash-test dummies for some anticipated AI of the future. The nay-sayers will claim that this is a deluded fantasy, that AI poses no threat to our future, and that any such claims cannot be based on quantitative real-world measurement. **But the effects of technology on our being have been so great that they defy measurement. Inter-generational gaps, the lack of social cohesion and the breakdown of communities: the consequences are so apparent and so pervasive that to grasp them in their entirety would be impossible.** To make such a measurement, we would probably need to build a supercomputer to account for all the behaviour patterns and social networks of every human on Earth. This in itself would be a hapless task, for the machine that does this may also be the one who nails the final hammer into our coffin. The recognition that there is an arena of conflict between man and machine must in sum make us realise once and for all that we must trust in ourselves before anything else. The widespread accumulation and adoption of technology has in the past few years shown its devastating aspect. Whether this has been the moment man and machine have been torn asunder and brought into Universal conflict, only time will tell.

20

Universality ★ Indeed, the Universal Subject does not centrally entail conflict, although our diagnosis of the time we are living in may imply as much. The most immediate conclusion that we reach in our positing of Universal subjectivity concerns our wider role and purpose as human beings. No longer are we drifting aimlessly around in an indifferent space, lacking subjectivity or even any sense of an objectively existing Universe outside ourselves. **Furthermore, the bifurcation of man and machine subjectivities announces to us through our technologically mediated being that the evolution of our consciousness is not simply correlated with the complexity of the machines we develop.** In today's world, the degree to which a society is technologically saturated may conversely indicate a malnourishment of the human senses. Universal subjectivity, in this sense, puts us back on track; once more we are in touch with the Universe, and above all, once more we are in touch with ourselves. The summit to where we are heading stands aloft again in the distance. We are not lost or trying in vain to reconstitute ourselves in the post-modern milieu of 'the death of the subject'. The Universal Subject carries with it endemically the notion that such a thing is possible, indeed that anything is possible. It is an eternal optimism in the capacity for humans to surpass themselves in the exploration and understanding of the Universe. As such, it is in tandem with human feats in science and engineering; in physics, chemistry and biology, indeed, all of the human sciences as well. Once it is established, not as a field of study, but as a particular outlook, then we may begin to sweep away the swathes of human thought that have been wrecked by academic post-modernism. **The empty spaces can be filled with new ideas, not born of overly critical pessimistic theory, but reflecting optimism in the future.** Such a thesis elevates science, technology and engineering, placing them above the lowly status given in the eyes of post-modern social constructivists. We can now see that science is socially constructed based on observations of an objectively existing Universe. Discussions on AI should also be invigorated. All-in-all, despite the disastrous scenario we are presented with, we reaffirm our faith in humanity's ability to reach a common goal. This view may hearken back to Enlightenment optimism, but in doing so, it could open the way to momentous progress for humanity in a time when such changes would otherwise seem impossible. Our main revelation here then may not have the Universal Subject as a goal we try to reach, but that presently, in our current state, Universal subjectivity is what we inhabit now. **We are Universal Subjects; it is only our task to realise it.** This is achieved in the time-honoured way of knowing one's place in the Universe. Apart from that, very little can be said, other than that we are at the beginning of a long journey, our destiny may even be held in space travel. But before then, there are many problems to be solved here on Earth before we even consider setting out to Mars and beyond. The visions we hold about our future tell us a great deal more about ourselves than anything else. Even so, the first step we take may determine all that follows; thus, in doing so, let us proclaim the following: **We are Universal Subjects!**

Afterword

An Optimist's Glass

For now we see through a glass, darkly; but then face to face...
The Holy Bible, Corinthians 13:12

The times we live in can be characterised by the challenges we face. The greatest of these has been described here, that of man's relationship with machines. The idea of the Universal Subject could also contain within it a kernel of wisdom: that perhaps man is the most dangerous machine of all. Most of the problems of today we have brought upon ourselves: environmental degradation, political strife and internecine conflict. The dangers that we face today are of our own making. This is the outlook that many people have of our condition and it must be respected. But to burrow one's head into the ground in an act of reflexive pessimism is not the answer. The eternal optimist in us tells us the opposite: today is the day of new beginnings, of new hopes and challenges to look forward to. We now have every resource at our disposal to meet these challenges head-on. Now that we have recognised and acknowledged our situation, never before in the history of man have we had such potential to shape our destiny; never before have we been so together and as one; never before has the world seen such an epoch of opportunity and potential advance. We stand on the brink of unlocking the gates to a most dazzling, promising and spectacular future. We stand on the threshold of a new era that will usher in breakthroughs and betterment. We stand having been summoned by a higher cause, that of the Universal Subject, that of ourselves and that of our very being. What elevates us is not that which we achieve through will alone, but that which we achieve through recognition, not just of ourselves on the way to the Universal, but of the other, the machine. Within our new framework, the progression of AI and of robotics will stand as a marker to our own development. At every step along the way, man and machine can thrive through mutual recognition. For this to happen, we must see the benefits machines and AI can bring us. If we see good in them, they might see good in us. In this sense, we reshape our future by envisioning what's best. Our perception of the world changes it, we can change something into its positive aspect just by looking at it a certain way. We can instil good in it through our perception alone. There are no barriers to what we can achieve. Even in the worst possible situation when all hope seems lost, there is a light at the end of the tunnel. We emerge through it and encounter some new scenery, where once again our faith in the world is restored. With father-time as our friend and luck on our side, the challenges of the future are to be welcomed, not shuddered at. With one push, humanity sets itself in motion again, with nothing to fear at all. With trepidation and curiosity, with renewed hope, with ourselves and with a love for life, we march onwards to the distant future. The time has come for us to dispose of the pessimism that has engendered the fallowing of intellectual life. Today, it is optimism that is the key to our very survival. Optimism unlocks the doors that seem to have locked up any path to progress and put us into stasis. Optimism is what can turn a seemingly lost and hopeless situation into a good outcome. Once we recognise this, then there is no cause for fearing the future, the present or indeed the past. It is said that the optimist sees the glass as being half-full, whilst the pessimist sees it as being half-empty. Well, if this is the case and we have chosen the path of optimism, then the Universal Subject must see the glass as being half-full.

So, let us now see through the glass, optimistically. Our hope in the stars, Our destiny.

Zero Books

CULTURE, SOCIETY & POLITICS

Contemporary culture has eliminated the concept and public figure of the intellectual. A cretinous anti-intellectualism presides, cheer-led by hacks in the pay of multinational corporations who reassure their bored readers that there is no need to rouse themselves from their stupor. Zer0 Books knows that another kind of discourse – intellectual without being academic, popular without being populist – is not only possible: it is already flourishing. Zer0 is convinced that in the unthinking, blandly consensual culture in which we live, critical and engaged theoretical reflection is more important than ever before.

If you have enjoyed this book, why not tell other readers by posting a review on your preferred book site.

Recent bestsellers from Zero Books are:

In the Dust of This Planet
Horror of Philosophy vol. 1
Eugene Thacker
In the first of a series of three books on the Horror of
Philosophy, *In the Dust of This Planet* offers the genre of horror
as a way of thinking about the unthinkable.
Paperback: 978-1-84694-676-9 ebook: 978-1-78099-010-1

Capitalist Realism
Is there no alternative?
Mark Fisher
An analysis of the ways in which capitalism has presented itself
as the only realistic political-economic system.
Paperback: 978-1-84694-317-1 ebook: 978-1-78099-734-6

Rebel Rebel
Chris O'Leary
David Bowie: every single song. Everything you want to know,
everything you didn't know.
Paperback: 978-1-78099-244-0 ebook: 978-1-78099-713-1

Cartographies of the Absolute
Alberto Toscano, Jeff Kinkle
An aesthetics of the economy for the twenty-first century.
Paperback: 978-1-78099-275-4 ebook: 978-1-78279-973-3

Malign Velocities
Accelerationism and Capitalism
Benjamin Noys
Long listed for the Bread and Roses Prize 2015, *Malign Velocities* argues against the need for speed, tracking acceleration as the symptom of the ongoing crises of capitalism.
Paperback: 978-1-78279-300-7 ebook: 978-1-78279-299-4

Meat Market
Female Flesh under Capitalism
Laurie Penny
A feminist dissection of women's bodies as the fleshy fulcrum of capitalist cannibalism, whereby women are both consumers and consumed.
Paperback: 978-1-84694-521-2 ebook: 978-1-84694-782-7

Poor but Sexy
Culture Clashes in Europe East and West
Agata Pyzik
How the East stayed East and the West stayed West.
Paperback: 978-1-78099-394-2 ebook: 978-1-78099-395-9

Romeo and Juliet in Palestine
Teaching Under Occupation
Tom Sperlinger
Life in the West Bank, the nature of pedagogy and the role of a university under occupation.
Paperback: 978-1-78279-637-4 ebook: 978-1-78279-636-7

Sweetening the Pill
or How we Got Hooked on Hormonal Birth Control
Holly Grigg-Spall
Has contraception liberated or oppressed women? *Sweetening the Pill* breaks the silence on the dark side of hormonal contraception.
Paperback: 978-1-78099-607-3 ebook: 978-1-78099-608-0

Why Are We The Good Guys?
Reclaiming your Mind from the Delusions of Propaganda
David Cromwell
A provocative challenge to the standard ideology that Western power is a benevolent force in the world.
Paperback: 978-1-78099-365-2 ebook: 978-1-78099-366-9

Readers of ebooks can buy or view any of these bestsellers by clicking on the live link in the title. Most titles are published in paperback and as an ebook. Paperbacks are available in traditional bookshops. Both print and ebook formats are available online.

Find more titles and sign up to our readers' newsletter at http://www.johnhuntpublishing.com/culture-and-politics

Follow us on Facebook
at https://www.facebook.com/ZeroBooks

and Twitter at https://twitter.com/Zer0Books